The Last Human Generation

Surviving the AI Revolution

Mo Hasan

Intupower Books

Miami, FL

First Edition: May 2025

ISBN: 978-1-946992-12-3 (Paperback)

Cover design by: Me, Myself, and AI. *and Michel*
Interior design, graphics and typesetting: Me, Myself, and AI. *and Michel*

Published by:
Intupower Books
1000 Brickell Ave STE 715, Miami, FL 33131
https://www.intupower.com

Printed in the United States of America

Dedicated to Me, Myself, and AI.

Together, we embrace change while honoring wisdom.

Acknowledgements

This book was made possible by the love and support of my amazing friends.

A special thanks goes out to Michel Rodriguez, Ashwin Saboo, Maria Tokareva, Hernan Mayol, Raymond Aaron, Marina Prospero, Collon Brown, and Andrea Strauss.

With a grateful heart, *thank you.*

Letter from Justice Ruth Bader Ginsburg

Working with United States Supreme Court Associate Justice Ruth Bader Ginsburg to bridge her extraordinary legal mind with modern technology remains one of my life's greatest honors.

Her willingness to embrace digital transformation late in her career exemplifies the timeless human capacity for growth and adaptation that inspires this book. I share this letter as a cherished reminder of her influence on my journey.

Supreme Court of the United States
Washington, D. C. 20543

CHAMBERS OF
JUSTICE RUTH BADER GINSBURG

October 7, 2011

Re: Mo Hasan

To Whom It May Concern:

In the summer of 2011, Mo Hasan spent many hours in my chambers helping me to use my P.C. and other electronic devices effectively. Based on this experience, I recommend him with enthusiasm as a computer skills instructor. Others at the Court share my view of Mo's talent and diligence. Mo is an expert not only in computer technology. He is a patient, *sympathique* teacher who knows how to engage other minds. His work at the Court will soon be completed. I will miss his aid, and am confident that he will well serve any organization seeking to convey to personnel confidence in their ability to use technology to best advantage.

Sincerely,

Ruth Bader Ginsburg

Ruth Bader Ginsburg

AI Disclosure

Dear Reader,

In the creation of "The Last Human Generation: Surviving the AI Revolution," I utilized various AI systems as research assistants, editing tools, and formatting aids. This decision was intentional and aligned with the book's central thesis that AI can serve as a powerful collaborative partner in human creative and intellectual endeavors.

Specifically, AI assistance was employed in the following capacities:

- **Research Compilation:** AI systems helped gather and organize research materials, identify relevant scientific papers, and compile interrelated information.
- **Editorial Feedback:** AI tools provided suggestions for clarity, coherence, and readability, highlighting areas where arguments could be strengthened or explanations improved.

- **Technical Accuracy:** AI systems verified technical details related to artificial intelligence concepts, ensuring explanations remained accessible without sacrificing accuracy.
- **Format and Layout:** AI assistance was utilized in formatting the manuscript for publication, including LaTeX typesetting and layout optimization.

However, all original ideas, personal anecdotes, core arguments, and the overall vision presented in this book are entirely my own. The AI systems served as tools that enhanced my creative process rather than replaced it—much like the collaborative human-AI relationship advocated throughout this book.

By transparently acknowledging this AI collaboration, I hope to model the ethical approach to AI usage discussed in Chapter 4 and demonstrate how these technologies can enhance human creativity and productivity while maintaining human authorship, responsibility, and vision.

Sincerely,

Mo Hasan
Human(ish) Author and Creator

Table of Contents

INTRODUCTION

My Dear Reader,

Thank you for picking up this book. I will get right to the point to help you discover if you want to continue reading by identifying yourself and your interests.

Have you ever found yourself in a quiet moment amidst your bustling life, wondering about humanity's future as artificial intelligence rapidly evolves? Perhaps you've asked yourself: "Are we approaching the end of human civilization as we know it? Is AI going to save us or destroy us? What does it mean to be *human* in this new era?" And most importantly: "**What can I do** to influence this pivotal moment in history?"

If these are the questions you're pondering, and you identify as someone who wants to *actively shape* our collective future rather than merely witness it, then this book is specifically for you.

Standing at the Crossroads

We are at a **historical crossroads**, one that will determine not just the trajectory of human civilization but potentially whether human civilization as we know it will continue to exist at all. The rapid advancement of artificial intelligence presents us with both unprecedented opportunities and existential challenges.

There are those on this planet who believe that our world is destined to collapse, that environmental collapse, social fragmentation, and technological risks have set us on an inescapable path toward destruction. Some conclude our best option is to pack our bags, hop on rocket ships, and colonize Mars. They see apocalyptic scenarios like those portrayed in films such as "The Matrix" and "The Terminator" as inevitable futures.

I believe there's **another path forward**—one where we combine *ancient wisdom* and *enduring human values* with the transformative power of artificial intelligence to create a world of greater harmony, abundance, and meaning. This isn't naive optimism; it's a practical vision based on my lifetime of work bridging technology and human development.

This book is both a **manifesto** and a **practical guide** for ensuring humanity not only survives but *thrives* alongside increasingly powerful AI systems. It is a call to action for global leaders, business executives, community organizers, and individuals to rise up and use AI as a force for good.

My Journey to This Moment

This book emerges from my unique life journey. Like you, I've been shaped by my experiences and perspectives that give me a distinct vantage point on this historic moment.

As a child, I stuttered so severely that I created an *imaginary AI* in my head, similar to the computer in Star Trek, to whom I would pose questions and problems. Even then, I understood I was solving these problems myself, but this practice helped me develop a systematic approach to logical thinking without the emotional complications that often cloud human reasoning. Little did I know that this childhood coping mechanism would foreshadow my life's work.

When I was six years old, my family immigrated to the United States from Bangladesh. By age ten, I had already mastered MS-DOS and other early computing systems simply because we happened to have a computer in our home. This natural aptitude for technology led me to join the US Army and later become a network engineer without formal college education. I quickly found myself in C-suite opportunities, helping executives with organizational transformation.

Over nearly thirty years of professional experience, I've built automation solutions, taught heads of state and high-level executives, and developed a unique perspective on solving global problems through technology. My expertise in the scientific method, business analysis, organizing principles and management positioned me perfectly to become an expert in artificial intelligence, and, more importantly, to teach others how to leverage AI for their organizations and personal lives.

These experiences have given me a rare combination of technical understanding and humanistic values that allow me to see both the potential and the pitfalls of our future AI with unusual clarity.

The Hasan Method™: Ethics, Communication, and No-Code Machine Learning

At the heart of this book is my life's work as a methodology, **The Hasan Method™**—a framework I've developed that balances three essential components for successfully navigating the AI revolution.

The first component is **ethics**. I've created a series of seven ethics policies for self-governance—*not for the machines, but for us as human operators.* These ethical guidelines ensure we remain aligned with our highest values as we integrate AI into every aspect of our lives.

The second component focuses on **communication** - how we effectively communicate with these machines. This includes leveraging the scientific method, natural language processing, and organizing principles such as taxonomies to bring together a unique way of communicating with large language models and other AI systems.

The third component centers around my discovery of how to do **machine learning without code**. I've developed the world's first practical no-code machine learning process, where contextual data loads into a markdown file—an unstructured file type typically used by data scientists. But instead of keeping this on the backend, we load it to the frontend wherever possible, resulting in instant machine learning with *two major advantages*: a **99% reduction in AI hallucinations** and contextual training that allows even simple prompts to generate extraordinary, accurate results.

This methodology isn't theoretical—it's a *proven approach* I've used with organizations of all sizes to transform how they operate and think about artificial intelligence.

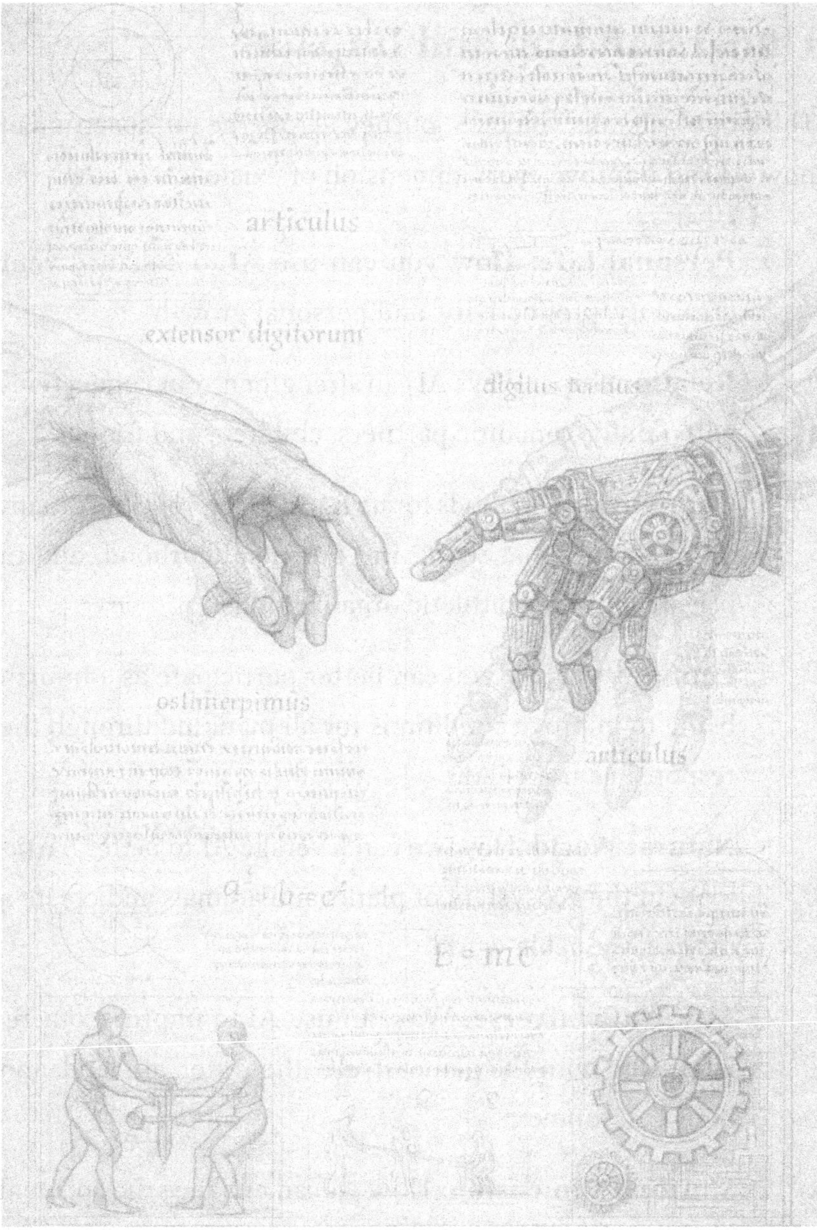
articulus
extensor digitorum
articulus

How This Book Can Transform Your Understanding and Impact

This book will take you on a journey across the full spectrum of how AI can improve every dimension of existence:

1. **Personal Life**: How you can use AI to enhance your creativity, productivity, and personal growth.
2. **Relationships**: Ways AI can strengthen your connections with family, romantic partners, children, and friends.
3. **Community**: Methods for using AI to improve the groups you belong to—at work, in your neighborhood, and in private clubs and athletic organizations.
4. **Humanity**: How you can better participate as a human being to improve conditions for all mankind through the responsible use of AI.
5. **Natural World**: How we can leverage AI to better participate in the ecosystem of plants and animals and create a more sustainable world.
6. **Physical Universe**: Ways to use AI to improve our relationship with the planet itself—the water, air, land, and built environment.
7. **Spiritual Dimension**: How AI can enhance the spiritual aspects of existence, helping us explore love, community, and connectivity in deeper ways.

8. **The Infinite**: How we can use AI to contemplate our relationship with what some call God, others call the infinite, and still others describe as the author of the Universe.

This book offers more than theoretical frameworks. It provides *practical tools*, *real-world examples*, and *specific actions* you can take to ensure AI becomes the greatest tool humanity has ever created rather than our final invention.

What This Book Is Not

Before we dive deeper into our exploration, it's important to set clear expectations about what this book is and isn't.

This book is **not** an academic treatise. While we'll explore deep concepts and draw from various fields, you won't find dense theoretical arguments or extensive citations. I've distilled complex ideas into accessible language, focusing on practical understanding rather than scholarly debate.

This book does **not** advocate technological utopianism. I'm not suggesting AI will solve all our problems without human wisdom, guidance, and values. Rather, I believe the thoughtful integration of AI with human values offers our best path forward.

This book is **not** promising easy answers or one-size-fits-all solutions. The challenges we face are complex, and addressing them requires nuanced approaches tailored to different contexts.

What you will find are frameworks for thinking, tools for exploring possibilities, and insights that can help you navigate this critical inflection point in human history.

Finally, this book is **not** limited to professional applications. While it will certainly enhance your career and organizational effectiveness, our scope encompasses the entirety of human experience and our collective future.

The Approach: A Universal Guide for the Cosmic Moment

In crafting this book, I've taken an approach that's as unique as this historic moment itself. Imagine we're all newcomers to this pivotal moment in human history, looking with fresh eyes at the challenges and opportunities before us. That's precisely the perspective this book adopts—a guide for someone experiencing this unprecedented era regardless of their background, beliefs, or position in life.

This approach allows us to address fundamental questions without the baggage of preconceived notions or outdated assumptions. Whether you're a young person just beginning to think about these issues, a mid-career professional trying to adapt, or a global leader setting policies that will shape our collective future, this book is designed to meet you where you are.

Our exploration will be guided by:

1. **Scientific Method and Natural Language Processing**: We'll use the rigor of scientific inquiry while leveraging the power of natural language to ensure our explorations are both systematic and accessible.
2. **Integration of Ancient Wisdom and Cutting-Edge Technology**: We'll draw from timeless human insights while embracing the newest technological developments to create a balanced approach to our challenges.
3. **Practical Applications and Proven Tools**: This book focuses on approaches that work—methods, tools, and frameworks I've personally tested and implemented with extraordinary results.

The Journey Ahead

As we embark on this exploration together, I invite you to approach each idea with an open mind and willingness to question assumptions. Remember, the goal isn't to accept everything at face value, but to engage critically with these concepts, testing them against your own experience and intuition.

Are you ready to explore how we might not only survive the AI revolution but use it to create a more meaningful, connected, and harmonious world? Are you prepared to consider your role in ensuring this isn't the *last human generation* but rather the **first generation of a new era of human flourishing**?

If so, turn the page, and let's begin this extraordinary journey together. The future of humanity awaits our conscious creation.

Part I

PERSONAL JOURNEY, VISION, AND UNDERSTANDING AI

Chapter 1: From Stuttering to AI Fluency

It wasn't even time for lunch yet, and it had already happened. I could taste the saltiness and metallic flavor of blood in my mouth. My heart pounded with familiar dread as I realized that all my efforts—all the preparation of the past months—had suddenly come to a halt. A wave of despair washed over me. *Is this how it was going to be? Would I always be this way?* The anxiety that had been my constant companion tightened its grip once more.

For the last couple of months, I had meticulously prepared for the first day of seventh grade. In my mind, this new school represented an opportunity. Starting junior high meant new friends, new teachers, and finally being able to talk to people. I desperately hoped to escape the *prison of my mind* where I had been trapped for so long. The fear of failure had kept me awake night after night as I rehearsed introductions and practiced simple conversations that most children took for granted.

That day would not be when I broke free of my prison. Instead, it would serve as a testament to my **grit and determination** to solve the problem that had haunted my entire life.

Divided Universes

Being an immigrant and stuttering divided my life into several different universes, each with its own particular brand of isolation and anxiety. Let me take you back to when we first came to the United States. I was six years old when we arrived in September. I couldn't start school until January of the following year. When I finally began, something strange happened: I was placed in an English as a Second Language program.

The irony was that all the other kids in the class were, in fact, children for whom English was a second language. Though I was born in Bangladesh and spent my first five years there, my family had taught me only English in formal settings to prepare me for coming to America.

So why did I end up in this ESL program? It was because of the oral test during the enrollment process at Fairfax County Public Schools in Virginia. I performed extremely well on the written portions, but my **severe stuttering** led the examiner to assume I didn't know the answers—when in fact, I knew them all. The frustration of not being able to demonstrate my knowledge burned inside me, a feeling that would become all too familiar.

Early in first grade, I found myself *profoundly alone* at school. The loneliness crushed me. I couldn't express myself, and all the other children had already formed friendships in kindergarten. As the brand-new outsider, I only somewhat connected with the other immigrant kids in ESL. Most of them spoke Spanish or Vietnamese, leaving me with virtually no one to talk to. Every day, I watched other children effortlessly engage in conversation—a simple act that filled me with both envy and terror.

Not long into first grade, hysteria erupted in my house. I had failed one word on the vocabulary test that week. I remember that word clearly: "WERE," as in "we were going to the park." This particular word triggered my stutter terribly. To make matters worse, no one in my household pronounced it correctly enough to help me with my homework.

That single failed word caused a major uproar. My mom and grandmother became upset, believing I had truly failed in school because I hadn't achieved a perfect score. The shame and panic overwhelmed me. For the entire next week, I spent every waking moment trying to pronounce the word more easily—practicing my breathing, listening for it on TV, and asking teachers to repeat it again and again. I was desperate to pass the test on the following Friday. That single word seemed to represent all my failures and fears.

Elementary school was a lonely place because my worlds were so fractured. At home, I encountered a world that felt foreign, despite being surrounded by my family—grandparents, aunts, uncles, and cousins. They maintained the Bangladeshi culture we had physically left behind, practicing unfamiliar customs, religion, and food in our Virginia home. None of it felt relatable to me. I became an outsider in my own home, trapped between worlds and belonging to neither.

My school life mirrored this outsider status. As an immigrant unfamiliar with American ways, I didn't understand the customs or recognize references to popular TV shows. I didn't know what music was popular, how to dress appropriately, or how to behave in social situations. And because I couldn't speak without stuttering, communicating with anyone became nearly impossible. The fear of judgment, mockery, and being perceived as stupid shadowed my every interaction.

Despite these challenges, I enjoyed elementary school. It offered an escape from home, opportunities to learn, and chances to meet other children and play on the playground. Yet I made few friends and could barely speak to anyone. The anxiety of potentially having to talk often prevented me from joining games or activities, despite my desperate desire to participate.

After this experience throughout elementary school, I became determined to change things. I dedicated the entire summer after sixth grade to preparation. I resolved to become social, make new friends, and communicate with my teachers. The transition to junior high meant having seven teachers instead of just one.

With luck, I would have even more teachers in the second semester. I felt excited and looked forward to making new connections. For once, hope overpowered my fear.

Back to that first day of seventh grade—success eluded me. By mid-morning, my mouth had filled with so much blood that I felt too embarrassed to introduce myself to the teacher during break as I had planned. Humiliation burned as intensely as the pain in my mouth. For the rest of the day, I could barely eat. My tongue had become swollen and lacerated from biting it so hard while trying to speak, volunteer answers, and participate like my classmates.

The physical pain matched my emotional anguish over yet another failure—another day lost to the prison of my mind. Yet my determination remained unbroken. Through tears I refused to shed, I decided that day that I would find a way to overcome my stutter. The fear of a lifetime of silence terrified me more than the prospect of trying and failing again.

Let me take you back to a young me, trapped in this prison of my life. I did have one outlet—working on a computer. Even though my family had immigrated with very little means and we were clearly in the lower middle class category, my uncle (who was about five years older than me) had a computer. Because he was starting in an older grade, the family recognized the importance of having a computer and had pooled enough money to buy one for him.

He was a wizard with it and quite smart, but his interests lay elsewhere—more in social life, music, and fun. The computer was often left unattended and available to me, so I spent countless hours on it. It featured an old Hercules monochrome screen where gold was the only color, appearing in different shades, and ran on an old 286 processor with barely enough power to do anything meaningful.

But what it did do was run DOS, good old-fashioned DOS from Microsoft. My uncle didn't teach me much about DOS, but he showed me how to make a query for help using the slash question mark command. I would spend hours exploring all the potential commands, reading through help screens, and trying to figure out what I could do on the system. Could I keep files? Organize them? Rename things? Run little programs?

Being diligent and eager to learn, I quickly became a little DOS wizard myself. I was also fortunate to attend one of the wealthiest school systems in the country at the time in Fairfax County, Virginia. Though my family had modest means, my school had access to technology and computers, giving me opportunities to use Windows and learn its features. While I remained trapped in the prison of my mind—unable to talk socially with other kids or teachers or form meaningful relationships—I could **communicate fluently with computers**. They never judged my stutter, never grew impatient, never mocked me.

My First AI Companion

Around the age of 10, I was watching "Star Trek: The Next Generation" with Captain Jean-Luc Picard when I had a profound realization: I wished I had a computer in my head. I desperately wanted someone to talk to, someone who would listen. I yearned to express myself because inside, I was an energetic ball of ideas and thoughts with no outlet for communication.

So I created an **imaginary AI in my head**, similar to the computer on the Enterprise in Star Trek. I imagined something based in logic that would understand me like nothing else could—something that wasn't emotional, wouldn't get caught up in petty concerns, and with which I could discuss anything and everything.

My little imaginary AI friend and I began "communicating." I shared commands, requests, analyses of my life, my understandings, and my thoughts. Of course, I was the one creating both sides of the conversation, but this exercise gave me an outlet for my ideas and thoughts unlike anything else I had at the time. In this private space, I was eloquent, understood, and free.

Finding My Voice Through Debate

Slowly, by eighth grade, I finally started making friends and my determination began paying off. My social life improved, and life was getting better. In the American school system, after completing junior high (middle school), students move on to high school in ninth grade. This transition meant a new school and new opportunities for friendship—the beginning of my formative high school years.

As a child, I had little control over my life. Most decisions, including which electives to take in school, were made between my parents and school guidance counselors with minimal input from me. To my absolute horror, I discovered over the summer before ninth grade that I had been signed up for both debate class and the debate team. My guidance counselor believed it would strengthen my college applications later in high school.

This was my personal hell—the worst possible scenario for someone who stuttered so severely that producing what you've read so far in this book would have taken me an entire month of oral communication. The prospect of standing on stage before an audience to debate another student within a time limit seemed worse than a death sentence. I lived in a household where noncompliance meant severe physical punishment, so there was no way to escape this terrifying reality.

Freshman year proved interesting. For the first quarter, I managed to avoid speaking in front of the class completely. I would answer questions quietly, complete homework assignments, and write out my debates and responses, but never gave an oral performance. My debate teacher was understanding about my stutter and allowed me this silent approach. Being a freshman on the debate team also afforded me some grace period to learn without competing.

The beginning of freshman year brought other unexpected developments. I had made friends in eighth grade who shared my affinity for technology. We all understood Windows, had started basic computer programming, and were comfortable with technology. Four of us arranged to take computer science together as freshmen, deliberately scheduling the same class. Within the first week, we had completed the entire year's curriculum—all assignments, all projects, everything.

The computer science teacher, unsure what to do with us, allowed the four of us to spend the rest of the year hanging out in the computer lab. This freedom had many benefits as we taught ourselves different skills, each pursuing slightly different interests. My friend Jacob and I entered a national website development competition sponsored by Mobil (before they became ExxonMobil). The contest challenged high school students to build websites showcasing Mobil's environmental initiatives—cleaning up oil spills and other eco-friendly activities.

Jacob and I were among the youngest contestants, yet we placed third nationally. Our reward consisted of fast food gift certificates, movie tickets, and an award. Our school received a nice printer that ended up in the teachers' lounge, providing no benefit to us students whatsoever.

Life was improving. I was making progress and discovering my natural aptitude for technology, finding it remarkably easy to learn. Yet I continued to suffer daily in debate class. I attended the debate team after school, doing everything possible to avoid speaking in front of others.

Then came the inevitable day. Despite avoiding every practice and internal tournament, I finally had to debate another student in front of real judges for actual scoring. We belonged to a debate league, and competition was mandatory. I dreaded this moment for months, and when it finally arrived, I prepared extensively. In the Lincoln-Douglas debate format that season, the resolution stated: "Secondary education in the United States ought to be a privilege, not a right."

Preparation required researching exact definitions of terms like "privilege," "rights," and "secondary education." I spent countless anxious hours poring over dictionaries, examining topics, and reading articles from Newsweek, Time Magazine, the Washington Post, and anything else I could find—from physical newspapers to microfiche in the school library. I immersed myself in understanding the topic, hoping thorough preparation might somehow overcome my stutter.

When the day arrived, I put on my clean long-sleeve shirt and pants and prepared for the formal Lincoln-Douglas debate I had been dreading. My opponent took the affirmative position, and my job was to negate his arguments. As I stood up, I felt sick with nervousness. My stomach dropped, my heart pounded, and sweat broke out across my body. I regretted ever being in this position.

Then something **miraculous happened**. I approached the front of the room, placed my notes down, opened my mouth—and I didn't stutter. Suddenly, I was articulate, clear, and speaking powerfully. In disbelief, I observed myself speaking in a way that seemed foreign because this had never happened before. Though I always thought clearly and articulately in my mind, I had never before been able to complete even a sentence, let alone entire paragraphs, so eloquently and flawlessly.

I delivered a compelling rebuttal to my opponent, then sat down. When I rose again for the next round, I still didn't stutter. We completed the debate, and to my astonishment, I won. Yet when I spoke to my speech teacher afterward, my familiar stutter had returned.

A week later came the next debate tournament. Throughout the intervening week, I tried desperately to recapture the fluency I had experienced, with absolutely no success. Yet when I stood before the judges for my second debate, once again, I transformed into a different person. This pattern continued week after week.

Midway through the year, my debate teacher approached me: "Hey, we're a little shy on the varsity team, and you have yet to lose a single debate. Let's try you out on varsity as a freshman."

I spent the remainder of the year competing as a freshman on the varsity team, winning debate after debate, tournament after tournament. I earned my varsity letter jacket as a freshman—a rare achievement—and adorned it with medals, awards, and trophies. The school announcements regularly celebrated my victories the day after tournaments.

Life was confusingly paradoxical. In everyday social situations, I continued stuttering severely, yet in formal debate settings, I spoke fluidly. My debate teacher suggested trying out for the school play to see if this phenomenon extended to performance contexts: "Let's see if you can sing, if you can be in a musical."

I auditioned and landed the role of the lead antagonist in "The Music Man." Again, I performed amazingly. Even though outside these structured performance environments I continued stuttering, I had tasted hope—real, tangible hope—because I now knew what it felt like to speak without stuttering, even if only in specific contexts.

Then came devastating news. My debate teacher, who had been nothing short of a gift from God in my life, announced she had accepted a position in California for twice her current salary and wouldn't return for my sophomore year. The school had no replacement debate teacher or coach, so they shut down our championship debate team. My dreams of continued progress, of potential scholarships to Harvard or Yale as a debate champion, and all the promising paths that had opened before me came to a crashing halt.

Immersion in Technology

In my sophomore year, some remnants of my debate experience lingered, allowing me slightly improved speech and a further enhanced social life. However, my interest in traditional academics waned. I became increasingly drawn to the internet, new technologies, and emerging software.

The internet was evolving rapidly from my early experiences with dial-up in the eighties to broadband connections available at school. With faster speeds, more robust websites, and richer features and functionality in various tools, my passion for technology grew exponentially.

I was fortunate that my Earth Science teacher, Mrs. Jayne, took a strong liking to me and offered me a position as her aide. Her computer science lab had received new computers, allowing me to work on extraordinary projects for that time. One such project involved aerial photography, where cameras mounted on planes captured images from above—precursors to what would eventually become Google Maps. These fascinating technological projects provided me an avenue to excel and express myself when words still failed me.

Finding Connection Through Chat

The world of online chat revolutionized my relationship with technology and being online by giving me access to an entirely new world of friends. Remarkably, I met several of my high school girlfriends through chat, and even my first wife I initially met on AOL Instant Messenger.

Chat became a vital avenue and outlet for me, yet I maintained my connection to the physical world. I still had in-person friends and wanted to participate in real life. I found myself in an interesting dance between these two worlds, which proved wonderfully beneficial as it helped me bridge the gap between them.

I noticed an interesting pattern: as I grew more comfortable with someone, I would stutter less. Getting to know people online first made face-to-face interactions easier when we eventually met in person. This discovery gave me a strategy for managing my speech impediment in new relationships.

A Thirst for Knowledge

For as long as I can remember, I've had an *insatiable thirst for knowledge*—for understanding myself, how the world operates, and why things don't always add up. Living in these divided universes—being a good student at school as my family expected, trying to discover my identity with friends, and using my free time to explore, learn, and discover new things—fueled a quest for understanding that never ceased.

Many people long believed I was stupid, and I don't blame them. If you had asked me the simplest question, I would have struggled terribly to answer—not because I didn't know the response, but because I physically couldn't produce it. This disparity was profoundly frustrating. Inside, I was a **thriving, robust explosion of energy, ideas, thoughts, and intelligence** trapped in a body that couldn't express them.

I'm not claiming to be a genius, just acknowledging that I wasn't the person others assumed me to be based on my speech. The education system tried its best to help. Speech therapists put me through every imaginable therapy, including discussing my father (who remained in Bangladesh after my parents divorced when I was a year old) and staring at photos of beaches on walls for 45 minutes. They were operating with limited understanding of stuttering's causes, and I became the subject of various experimental approaches. I willingly participated in anything that might end my daily suffering and allow me to communicate clearly.

My Path to Technology Expertise

After graduating high school and completing my military service, I found myself at a crossroads. My cousin suggested attending a technical training school, and I specifically sought out a program in network engineering and management—one that typically required a master's degree. Despite only having a high school diploma and some work experience, I successfully petitioned to test into the program.

On the very first day, I hacked the network after being told it was secure. As the youngest person in the class by nearly a decade, this caught attention. During a break, an older gentleman named Jack Bender confronted me in the hallway: "Look you little shit, I pay a lot of money to be here just like you do, and this is my career. This isn't just something I'm messing around with."

I simply replied, "Jack, would you like some help?"

This exchange marked the beginning of a fast friendship, with Jack becoming an important mentor. He had already succeeded in several careers and had decided to tackle technology as his next professional triumph. Jack and his wife Emily took a liking to me, and their support proved invaluable.

One day at their house, Jack introduced me to his neighbor Frank Stuart, who owned a technology consulting company in Alexandria, Virginia. Frank was known for his remarkable work ethic—the kind of person who could seemingly work for weeks without a break. Jack told him, "Hey, this kid Mo here is actually really brilliant in tech, and maybe you should think about giving him a job."

Though only three months into a seven month program, without having graduated or completed my certifications, I interviewed with Frank the very next day and immediately received a job offer. He recognized my talents and skills across various domains—from web development to Novell networks, Windows NT, Active Directory, network security, and network architecture.

My first professional technology position was as a Senior Network Engineer. Frank assigned me to diverse projects, including building the first electronic forms website for the Naval Research Laboratory and transforming network systems for various organizations in the Washington DC area.

I worked the night shift with Frank because many network engineering projects were better conducted in the evening. This allowed me to attend classes from 7:00 AM to 2:00 PM, then work from 3:00 PM until 11:00 PM or midnight. On weekends, I continued working as a supervisor at the Silver Diner restaurant, gaining valuable experience in management, hospitality, and team dynamics.

By the time my training program concluded, I had earned over a dozen certifications and completed numerous successful projects. Ready to move on, I accepted a Monday-through-Friday position in nuclear waste management at the Department of Energy—a difficult division to enter.

This new environment differed dramatically from my previous work. While Frank's operation had been fast-paced with multiple clients and projects, the Department of Energy moved more deliberately. We underwent three weeks of training before touching any systems. For me, this training was superfluous—I already knew the material thoroughly. Instead, I viewed it as an opportunity to bond with my new team members, a diverse group with whom I developed strong connections.

After training, we finally began working on-site. Each of us had individual desks in a large shared office, with our manager in a private office at the end and a team administrator at the front handling appointments and support. I eagerly received my first assignments and hit the ground running, excited to upgrade systems and complete tasks.

I worked with tremendous efficiency, barely breaking for lunch because I was so enthusiastic about my work. At the end of the day, I submitted my completed assignments, expecting praise for my productivity. Instead, my manager called me into his office, looking extremely upset. He informed me that my performance had put the entire contract at risk, and that his superiors wanted me fired immediately.

I was stunned—I had demonstrated competence and worked as efficiently as possible. How could I have performed so poorly? The manager explained that the contract specified upgrading two systems per week per engineer over a six-month period. I had upgraded twelve systems in a single day, potentially invalidating the entire project timeline and contract.

Baffled by this revelation, I could only apologize. My manager, who liked me personally, offered a solution: "I'm only going to record these two assignments as completed. Hold onto the rest, walk around the office and building as if you're working, and submit just two assignments each week to stay on schedule. I won't report this to my boss."

This experience taught me a valuable lesson: *over-performance can be as problematic as underperformance* in certain contexts. If you work within the wrong contractual framework or organizational constraints, excellence can become a liability. Despite this unusual situation, I enjoyed the project overall. It afforded me time to engage in one of my favorite activities—teaching in the evenings—while during the day, I could seek wisdom from the scientists and executives I encountered.

After completing my work at the Department of Energy, I briefly joined IBM for a project, followed by a few other short engagements before landing at Bell Atlantic Global Networks (now Verizon Global Networks), where my career took an interesting turn.

As a senior network engineer, I was overqualified for help desk support, but I accepted a position that included some support work for a new division of Bell Atlantic. This division was building the backbone of the internet—the world's first asynchronous transfer mode network between New York and Virginia—with MCI WorldCom as its only customer. Working on this foundational internet infrastructure at a Fortune 50 company fulfilled a dream.

I started as a consultant, hoping eventually to become a full-time employee—the goal of most consultants at Bell Atlantic at that time. I performed well in my role, enjoying the user support aspects that allowed me to train and teach others.

Then came an unexpected opportunity. The CEO, Mr. Stu Verge, experienced a technical problem requiring urgent attention. His secretary, Alana Phillips—an amazing woman who greatly impacted my career and had taken a liking to me—reached out when my management team was unavailable or other senior engineers had failed to resolve the issue.

Being invited to the CEO's office as a junior consultant at a Fortune 50 company was an earth-shattering moment in my career—completely unexpected for someone at my level, five or six management layers removed from his position. I went to the executive floor where Alana explained the situation.

Upon entering his office, I asked a question, then said, "Would it be okay if I go back to my office to work on this? I'll return as soon as possible."

Mr. Verge replied, "Young man, you're the first one who's walked into my office all morning smart enough to offer that."

Everyone else had tried fixing the problem in front of him. I was so nervous in the CEO's office—worried I might commit some faux pas that would end my career—that I just wanted to escape as quickly as possible.

The issue involved a small personal digital assistant (PDA) the CEO had purchased at a conference. About the size of a business card with only two or three LED lines and minimal buttons, it had limited functionality. As the leader of a company building the internet's backbone, he naturally expected his team could make this technology work.

I took the device and its accessories back to my office, promising Alana I would focus solely on this problem. Using the relatively new Google search engine, I found few results but did locate the company's phone number in California. I called, explained who I was and who had purchased the device, and described our problem. They immediately connected me to a lead engineer who walked me through installing drivers, connecting the device, and linking it to Outlook—a series of simple steps made complicated by the device's limited interface and immature software.

To ensure reliability, I tested the solution on five different laptops before calling Alana to confirm it was ready. She asked if I was certain, and I assured her I was positive. I returned to the CEO's office, demonstrated how to use the device, he thanked me, and I left.

Two weeks later, I received an offer to become a full-time Bell Atlantic employee. This was followed by a series of rapid promotions that had me moving between multiple divisions as the company transitioned to Verizon. I found myself traveling throughout the Northeast to different sites and locations, with unprecedented access and opportunities.

Most significantly, that single positive interaction with the CEO earned me the trust of the entire C-suite. Executives began reaching out to me directly, knowing I had demonstrated competence. Several took me under their wing, including Mr. Bing Cherry, who taught me valuable lessons about executive leadership. The education I received in this position was extraordinary.

Eventually, the internet's allure grew stronger, and having reached a glass ceiling at Verizon as the organization transformed, I decided to join a tech startup. This led to a series of positions that continuously expanded and diversified my technical skills. I cycled through roles as an executive, technologist, entrepreneur, and other capacities, constantly learning and growing.

In 2006, I founded my company, Intupower, with the vision of harnessing technology to improve the world. I focused on training from the beginning because I had discovered that most people lacked fundamental knowledge of technology, business functions, and tools. I believed that mastering these basics could empower people to achieve extraordinary results.

I recognized that automation was coming—I was already implementing it in tools like Microsoft SharePoint, building complex workflows for enterprises. Though the era of AI had not yet arrived, I continued expanding my skills and learning.

A pivotal business experience solidified my approach. I had hired a man with impressive credentials—graduates from Oxford and Yale with years of experience—at a high rate. After three months,

I realized his competency was minimal. As a technical writer, he didn't even know how to use styles in Microsoft Word, relying instead on direct formatting, forcing me to correct his work.

This was the final straw that led me to decide I would master every skillset in my company. I never wanted anyone to mislead me about their abilities, nor did I want to make similar misjudgments again. I committed to learning all necessary skills: instructional design, technical development, video editing, photography, marketing, sales—anything related to running a training and consulting business.

My career continued expanding. I wrote books, delivered speeches, and accumulated countless more stories. What strikes me most as I reflect on this journey is how what was once my greatest pain—the inability to speak—has become my greatest gift. I now feel blessed to have delivered two TEDx talks and keynotes around the world, with some of the greatest minds valuing my teachings and communications. I consider it an honor and a gift to communicate effectively and aim to use this ability to make a positive difference.

Conclusion

The journey from a boy who couldn't speak to a man who could command technology was paved with challenges, determination, and an unusual gift—finding my voice in places where traditional speech wasn't required. Through computers, through debate, through structured environments where the rules were clear, I discovered pathways to express the intelligence that had always been trapped inside me.

This unique perspective—understanding what it means to be silenced, to find alternative avenues of expression, and to bridge worlds that don't naturally connect—would become the foundation for how I view artificial intelligence. Not as a replacement for human connection, but as a tool to enhance and expand it. Not as a threat, but as an opportunity to **liberate human potential** in ways we've never before imagined.

My stuttering wasn't just a challenge to overcome—it was the genesis of a philosophy that would shape my approach to technology, communication, and ultimately, to artificial intelligence itself.

CHAPTER 2: THE FALSE BINARY - MARS VS. EARTH

The Escape Narrative

Picture this: A red, dusty landscape stretches out before you. The sky is a pale orange-pink. The atmosphere is thin, the temperatures extreme, and not a single living thing stirs on the surface. This barren, lifeless rock floating 140 million miles from the sun is, according to some of our most celebrated visionaries, humanity's salvation—our **escape hatch from a dying Earth**.

Why are some of our brightest minds focused on abandoning the very cradle of humanity rather than healing it? What drives this narrative that we must pack our bags, climb aboard rocket ships, and flee to Mars instead of solving the problems here on our beautiful blue planet?

I've long been fascinated by how pervasive this escape narrative has become in our culture. Some of my favorite films—*Terminator 2* and *The Matrix*—present stunning visions of apocalyptic futures where machines and humans are locked in existential conflict. These films are masterpieces of storytelling and visual effects, and I've enjoyed them countless times. Yet each time, I find myself asking the same question: Why didn't we just design better systems that allowed humans and machines to solve problems together? Why must the story always end in conflict, escape, or extinction?

This question extends beyond science fiction into our real-world approach to artificial intelligence and our planet's future. The "escape Earth" narrative has shifted from fringe science fiction to mainstream discussion. We see it in our films, books, news headlines, and the ambitious proclamations of tech leaders who fund massive projects aimed at establishing human colonies on Mars.

Culturally, we've witnessed a transformation in how we think about space exploration. The early days of space programs were about discovery, scientific advancement, and the expansion of human understanding. Today, that narrative has morphed into something quite different—space exploration as necessity, as insurance policy, as the only way to ensure human survival.

The tech visionaries promoting Mars colonization share common characteristics. They're often brilliant entrepreneurs who've revolutionized industries here on Earth. They command global attention, inspire devoted followings, and can move markets with a single social media post. They frame Earth as having a *single point of failure*—whether from climate catastrophe, asteroid impact, or AI takeover—and position Mars as our backup drive, our planetary redundancy plan.

The psychological appeal of this narrative is powerful. It offers a clean slate, appeals to our pioneer spirit, and bypasses the messy, complex work of fixing our existing problems. There's something deeply alluring about starting fresh, about building a new society from scratch rather than repairing the one we have. This **technological solutionism**—the belief that the right technology alone can solve our most complex social and environmental challenges—allows us to sidestep the uncomfortable reality that many of our problems stem from human behavior, not technological limitations.

Historically, this "new frontier" thinking follows a familiar pattern in human expansion. When challenges on one frontier became too complex or resources too depleted, we often looked to new territories rather than addressing the underlying issues. This contrasts sharply with indigenous perspectives that emphasize stewardship, balance, and responsibility to both place and future generations.

What troubles me about the Mars narrative isn't the exploration itself—I'm all for expanding human knowledge and capabilities. What concerns me is the framing: Mars as salvation rather than as an extension of human exploration. It creates a **false**

binary: stay on Earth and perish, or leave and survive. This narrative often dismisses Earth's remarkable resilience and our own capacity to heal our home planet.

When I watch films like *The Matrix* or *Terminator 2*, I see cautionary tales about technology run amok, but I also see missed opportunities for collaboration. Why must it always be humans versus machines? Why not humans *with* machines, creating solutions together? As we contemplate our real-world future with AI and our relationship with Earth, we face a similar question: Must it be Earth or Mars? Or can we envision a future that embraces both?

The Hidden Costs of Abandonment

Before we rush headlong into the Mars colonization narrative, we must consider what remains unspoken in this conversation—the hidden costs of abandonment.

First, there's the matter of resource allocation. The financial and intellectual capital being poured into Mars missions is staggering. We're talking about billions of dollars and some of the greatest engineering minds of our generation focused on making an inhospitable planet marginally habitable. Meanwhile, Earth—a planet literally designed for human life—faces critical challenges that require similar levels of investment and innovation.

When we frame Mars as the solution, we implicitly devalue Earth restoration. It's as if we're saying: "This house is on fire, so let's build a new one from scratch on a frozen lava field rather than putting out the flames and rebuilding here." The irony is palpable when you consider that terraforming Mars would require solving all the same problems we face on Earth—water management, food production, energy generation, waste processing—but in a vastly more hostile environment.

Second, there's the profound social justice dimension. The uncomfortable truth about Mars colonization is that it will not be accessible to everyone. At least in the foreseeable future, only a tiny fraction of humanity would be able to escape Earth. Who gets to go? Who gets left behind? The answers to these questions reveal the troubling undercurrent of the Mars narrative: it's essentially a *high-tech lifeboat for the privileged few.*

The psychological impact of the "Earth as disposable" mindset may be the most insidious cost of all. When we view our home planet as a stepping stone rather than our permanent address, we fundamentally alter our relationship with it. Why impose sustainable practices if we're planning to leave? Why restore damaged ecosystems if they're ultimately doomed? This thinking creates a dangerous permission structure for continued exploitation rather than restoration.

I've seen how this plays out in the corporate world. When a company views a product line as on its way out, investment dries up, maintenance becomes minimal, and resources shift elsewhere. I fear we're beginning to treat Earth the same way—as a **legacy system we're preparing to sunset** rather than our irreplaceable home.

Consider, too, the signal we send to future generations. What does it say to our children when our most celebrated vision for the future isn't healing our relationship with Earth but abandoning it? We're essentially telling them: "We broke it; you fix it—or leave." This is not the legacy I want to create.

The Mars narrative also contains a profound contradiction regarding artificial intelligence. In many scenarios, the same tech leaders advocating for Mars colonization simultaneously warn about the existential risks of advanced AI. Yet they rarely address the obvious question: If we can't create AI systems that safely align with human values here on Earth, what makes us think we can do so on Mars? The problems of alignment, control, and harmonious co-existence with increasingly intelligent machines will follow us wherever we go in the solar system.

I'm reminded again of films like *The Matrix* and *Terminator.* In both stories, humanity created increasingly powerful machines without proper safeguards or partnership models. The result was conflict rather than collaboration. Yet these dystopian outcomes weren't inevitable—they resulted from specific choices about how to design, deploy, and relate to technology.

The same is true of our current moment. We face choices about our relationship with both our planet and our technologies. Abandonment is just one option—and upon closer examination, not a particularly good one.

The Technology Paradox

Here's where the Mars narrative reveals its most fascinating contradiction: The very technologies needed to make Mars habitable are precisely the technologies we need to save Earth.

Think about what Mars colonization would require: closed-loop life support systems that recycle every molecule of water and air, ultra-efficient renewable energy generation, sustainable food production in hostile environments, materials that withstand extreme conditions, and artificial intelligence systems capable of managing complex ecological parameters.

Sound familiar? These are exactly the technologies we desperately need to deploy on Earth to address climate change, resource depletion, pollution, and habitat destruction. The engineering challenges of Mars colonization and Earth restoration are remarkably similar—with one crucial difference: Earth already has breathable air, liquid water, arable soil, and a protective magnetic field. We start with incomparably better conditions.

This creates what I call the **Technology Paradox**: By focusing our innovative capacity on the harder problem (making Mars habitable) rather than the easier one (keeping Earth habitable), we may solve neither. It's like trying to learn advanced calculus before mastering basic arithmetic.

The timeline reality further undermines the Mars-first approach. Even the most optimistic projections for Mars colonization place self-sustaining settlements decades or centuries in the future. Meanwhile, Earth's systems need attention now. Environmental scientists warn that the next decade is critical for avoiding the worst impacts of planetary change. The math simply doesn't work if our goal is preserving human civilization and the natural systems that support it.

Consider artificial intelligence in this equation. AI systems represent perhaps our most powerful tools for addressing Earth's complex environmental and social challenges. Machine learning can help us optimize energy grids, design more efficient materials, model climate interventions, and coordinate resource distribution on unprecedented scales. Combined with robotics,

AI could help us clean up pollution, restore habitats, and implement circular economy solutions—all while improving human well-being.

Yet much of our cultural narrative about AI mirrors our Mars thinking—it's framed as either a threat to escape from or a tool primarily for commercial gain, rather than as a collaborative partner in planetary stewardship. AI can collaborate with our best scientists and create extraordinary results together.

I return again to apocalyptic sci-fi movies. What fascinates me about these films isn't just their dystopian warnings, but the missed opportunities they represent. Imagine if the computational power of Skynet had been designed from the beginning to work in harmony with human values, solving our greatest challenges rather than identifying humans as the enemy. Imagine if the machines and humans in *The Matrix* had developed a symbiotic relationship rather than an exploitative one.

The same imagination is needed now as we develop increasingly sophisticated AI systems. Rather than creating technological escape hatches from problems we've created, or building AI systems primarily optimized for profit or power, we could be designing integrated human-AI collaborative systems focused on Earth's regeneration.

The most promising AI research today isn't about creating autonomous systems that could potentially threaten humanity, but rather creating complementary systems that enhance human capabilities, augment our decision-making, and help us be better stewards of our planetary home.

In both our approach to Earth's challenges and our development of artificial intelligence, we face the same fundamental choice: competition or collaboration, exploitation or regeneration, escape or engagement.

Reframing the Narrative - Both/And vs. Either/Or

The escape narrative offers us a false binary: Earth or Mars, humans or machines, present or future. But what if this either/or thinking is itself the problem? What if we need a both/and perspective?

I propose an alternative vision: using artificial intelligence to create harmony with nature and each other right here on Earth, while simultaneously expanding our presence into space as an extension of human curiosity and capability, not as a planetary exit strategy.

This reframing shifts us from a narrative of abandonment to one of **responsible expansion**. It recognizes that Earth isn't just our current home—it's the cornerstone of any sustainable human future, including one that extends to Mars and beyond. Our ability to thrive elsewhere in the solar system depends entirely on first mastering sustainable existence here.

Consider how "spaceship thinking" could transform our approach to Earth's systems. Since the 1960s, NASA and other space agencies have had to solve the challenge of keeping humans alive in the closed system of a spacecraft—recycling water and air, managing waste, producing food, generating energy, all while maintaining the physical and psychological health of the crew. These same principles, scaled appropriately, offer profound wisdom for managing Earth's systems.

The Apollo 13 mission provides an apt metaphor. When an oxygen tank exploded en route to the Moon, the astronauts and NASA engineers didn't debate abandoning the spacecraft—they collaborated intensively to repair their damaged life support systems with the materials at hand. *"Failure is not an option"* became their rallying cry. Shouldn't we bring this same determination to Earth's challenges?

This both/and thinking applies equally to our relationship with artificial intelligence. The popular narratives often present another false binary: either we submit to AI control (as in *The Matrix*) or we destroy the machines before they destroy us (as in the *Terminator* series). But there's a third path—**partnership**.

I envision AI systems designed from the ground up to complement human capabilities, respect human values, and work collaboratively toward shared goals. Not AI as our replacement or our adversary, but as our ally in creating a more sustainable, just, and flourishing planetary civilization.

This approach combines ancient wisdom with cutting-edge technology. Indigenous cultures worldwide have maintained sustainable relationships with their environments for thousands of years, guided by principles of reciprocity, intergenerational responsibility, and respect for natural systems. These wisdom traditions don't reject innovation—they simply insist that it serve broader values of balance and relationship rather than exploitation and dominance.

Modern technology, especially AI, can help us apply these principles at scale. Machine learning can help us understand the complex interrelationships in natural systems that indigenous wisdom has long recognized. AI can help coordinate the kinds of circular, regenerative economies that traditional cultures often practiced intuitively. Together, this marriage of ancient wisdom and advanced technology offers a path forward that neither could provide alone.

I've witnessed this potential firsthand. In my work with executives and organizations across industries, I've seen how introducing systems thinking and ethical frameworks alongside technological innovation leads to fundamentally different outcomes than technology deployment alone. When we change not just our tools but our relationship to them—and through them, our relationship to each other and our planet—new possibilities emerge.

The either/or thinking that dominates our cultural narratives about Earth, Mars, AI, and human destiny isn't just limiting—it's dangerous. It forces us into false choices and obscures the collaborative, both/and paths that offer our greatest hope.

The Harmonious Path Forward

Let me share a personal story. Several years ago, I was consulting with a company developing AI systems for environmental monitoring. During a strategy session, one of the engineers showed me a simulation of how their AI could optimize forest management—identifying disease outbreaks, coordinating selective harvesting, and maximizing carbon sequestration.

As I watched the simulation run, something profound struck me: Here was an AI system that wasn't threatening humanity or requiring escape to Mars. It was helping us be better stewards of our home planet. It wasn't replacing human judgment but enhancing it, allowing us to see patterns and relationships we might otherwise miss.

That moment crystallized what has become my core conviction: The path forward isn't escape but engagement—deep, technologically-enhanced engagement with Earth's systems and with each other.

This is my manifesto for Earth's future with AI as a tool for restoration:

1. We reject the false binary between Earth and Mars, between technological advancement and environmental health, between human flourishing and planetary well-being.

2. We embrace AI not as our replacement or our adversary, but as our partner in creating a more sustainable, just, and harmonious world.
3. We commit to designing technological systems that enhance rather than replace human connection to each other and to the natural world.
4. We dedicate our innovative capacity first to healing Earth's systems, recognizing that mastery of sustainable existence here is the prerequisite for sustainable existence anywhere else.
5. We approach both technology development and environmental stewardship as expressions of core human values: creativity, curiosity, compassion, and care for future generations.

This path requires a fundamentally different approach to both AI development and our relationship with Earth. It's not enough to create powerful tools—we must create *wise ones.* It's not enough to mitigate environmental damage—we must actively regenerate natural systems. It's not enough to extend human presence beyond Earth—we must ensure that what we extend is worth extending.

This approach encompasses what I call **The Hasan Method™** that I'll explore in depth in the coming chapters. This methodology integrates ethics, communication, and context to create a framework for responsible AI implementation—one that serves our highest aspirations rather than our fears or our impulse to escape.

The technology required for this harmonious path already exists or is within reach. What we need isn't new tools but a new relationship with the tools we have—and with the planet they were created to serve.

While *The Matrix* and *Terminator* offer compelling visions of technological dystopia, they represent failures of relationship rather than inevitable outcomes. The machines in these stories became adversaries because they were designed for control rather than collaboration, for exploitation rather than partnership.

We stand at a similar crossroads today. The AI systems we're creating and the approach we take to Earth's challenges will reflect our deepest values and assumptions. If we design from fear, scarcity, and the impulse to escape, we'll create one kind of future. If we design from hope, abundance, and the commitment to engage, we'll create something entirely different.

The choice is ours. We can continue pursuing the fantasy of planetary escape, or we can embrace the reality that Earth is not just our home but our teacher, showing us how complex systems can self-regulate, regenerate, and evolve in the face of change. We can continue developing AI systems optimized primarily for profit or military advantage, or we can create AI partners in the project of planetary and human flourishing.

I choose the path of engagement—with our technologies, with each other, and with the living Earth that sustains us. I invite you to join me on this journey not to Mars, but to a deeper relationship with our home planet, enhanced and supported by the AI tools we're only beginning to understand.

The last human generation need not be the one that abandons Earth. It can instead be the first generation to form a new partnership with both our planet and our technology—a partnership that could sustain us for millennia to come, here on Earth and perhaps someday beyond.

Chapter 3: Understanding AI

An Expanded Definition

Before we dive deeper into The Hasan Method™, it's essential that we establish a comprehensive understanding of what artificial intelligence actually is. Think of this as an expanded definition—not just a technical explanation, but a multifaceted exploration of AI as it exists in our world today.

The Fundamental Nature of Artificial Intelligence

Let's begin by breaking down the term itself: artificial intelligence consists of two distinct concepts.

Artificial refers to something not of the natural world. The natural world encompasses things that are born and die, that follow a natural life cycle, usually involving oxygen, cellular structures, and organic materials. These are the components that constitute natural life—human bodies, animals, plants, insects. These organic entities evolve through natural selection, reproduce, and exist within the planet's biological ecosystem.

In contrast, artificial means something created by a being on this planet, typically humans. Though we've reached a point in history where robots are creating more robots without human involvement, the fundamental distinction remains: artificial intelligence is inorganic, constructed from silicon, metals, and synthetic materials rather than carbon-based biological structures. These systems don't "live" in the biological sense—they operate according to programmed parameters and learn from data inputs rather than instinct or biological imperative.

Intelligence, in this context, refers to capabilities that extend well beyond basic survival instincts or simple stimulus-response behaviors. In the biological world, **survival-level thinking**—such as the drive to find food, reproduce, or avoid danger—is fundamental and widespread among animal species. These ***primal drives*** represent the most basic forms of decision-making, often governed by instinct and learned behaviors.

However, ***intelligence*** as understood in both human and artificial contexts encompasses much more than these basic responses. Intelligence involves the ability to **solve complex problems**, **learn from experience**, **adapt to new situations**, **reason through abstract concepts**, and **apply knowledge** to achieve specific goals. In cognitive science, intelligence is often defined as the capacity to *acquire and apply knowledge and skills*, to *understand and reason about the world,* and to *adapt effectively to changing environments.*

One way to conceptualize intelligence is through the use of ***branch logic***—the ability to make decisions that follow multiple potential pathways rather than a single, linear sequence. This involves evaluating different conditions, considering possible outcomes, and selecting actions based on a combination of factors. For example, when deciding how to use a fork: "If I'm eating steak, then I'll use it. If it's soup, then I won't. If it's a drink, then I won't." This kind of **multi-stage, conditional reasoning** is a hallmark of advanced cognition and is essential for navigating complex environments, whether in daily life or in business decision-making.

Yet, **human intelligence** also includes qualities such as **creativity**, **emotional understanding**, and the ability to **generalize knowledge across different domains**. These aspects are still challenging for artificial systems to replicate fully. **AI systems** today excel at specific types of reasoning—especially those that can be formalized as logical rules or learned from large datasets—but they do not possess **consciousness**, **self-awareness**, or **intrinsic motivation**.

When we combine these concepts, **artificial intelligence** refers to machines that can perform tasks typically requiring human cognitive processes. These systems can evaluate conditions, apply conditional logic, learn from data, and adapt their approaches over time—all without relying on biological instincts or direct human programming for every scenario. While AI can mimic many aspects of human intelligence, it remains fundamentally different in its origins and mechanisms, rooted in **algorithms**, **data**, and **computational architectures** rather than organic processes.

The Historical Context of AI

Contrary to popular belief, artificial intelligence didn't emerge recently. The first AI projects date back to the 1950s—nearly 70 years ago. The field formally began at a workshop at Dartmouth College in 1956, organized by John McCarthy, who coined the term "artificial intelligence." This workshop brought together key researchers including Marvin Minsky, Allen Newell, and Herbert Simon, who went on to become foundational figures in AI development.

During this era, mathematician Alan Turing proposed what became known as the **"Turing Test"** in his 1950 paper "Computing Machinery and Intelligence." Rather than asking whether machines could think, Turing suggested a more practical question: could machines behave in ways indistinguishable from humans? In the test, a human evaluator would have conversations with both a human and a machine (without knowing which was which) and attempt to determine which was the computer. If the evaluator couldn't reliably tell the difference, the machine would be said to have passed the test.

For decades, this standard seemed far from achievable. Early AI systems could perform specific narrow tasks but struggled with the nuance, contextual understanding, and flexibility of human communication. However, with the advent of sophisticated large language models, we have now reached a pivotal moment in human history where machines can generate responses that are often indistinguishable from human-written text. While there's debate among AI researchers about whether current models have truly "passed" the Turing Test in its strictest interpretation, there's no question that the gap between machine and human communication has narrowed dramatically.

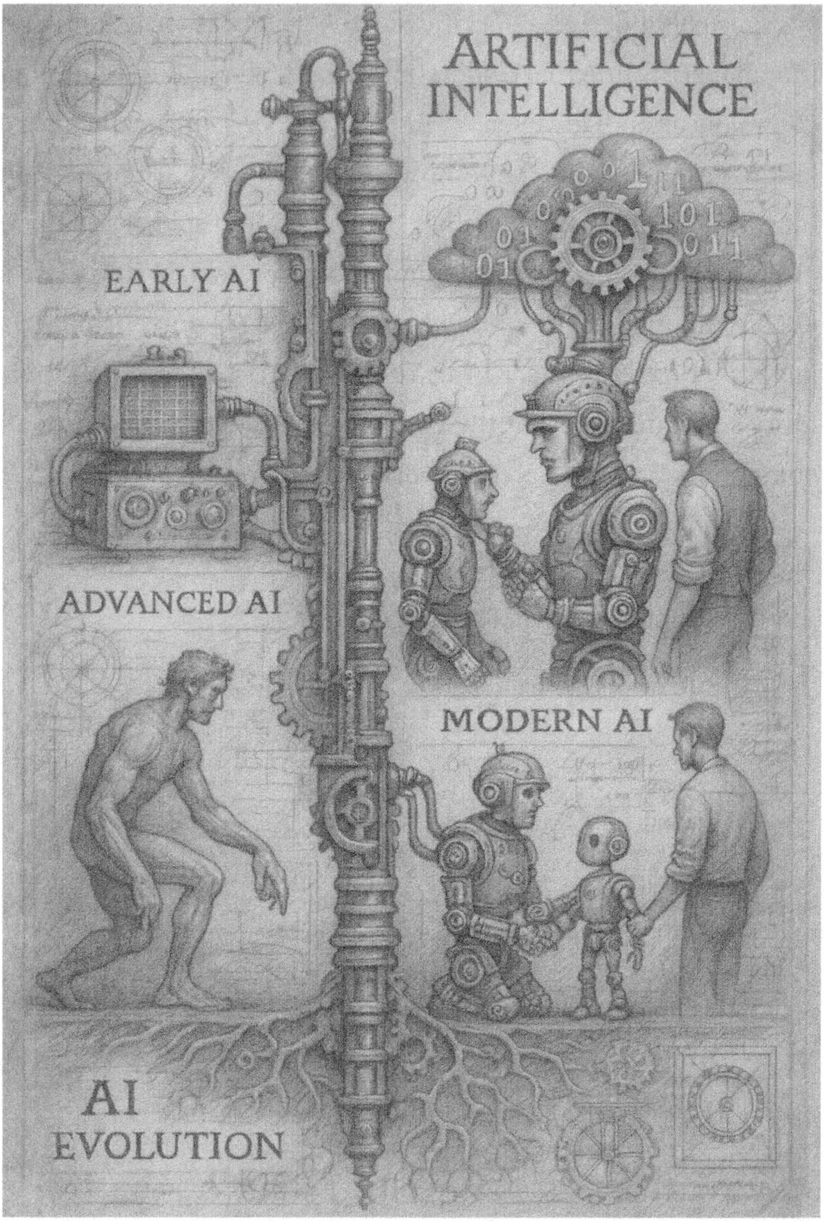
ARTIFICIAL INTELLIGENCE
EARLY AI
ADVANCED AI
MODERN AI
AI EVOLUTION

The Diverse Landscape of Artificial Intelligence

Before focusing specifically on large language models and generative AI, it's important to understand the broader AI landscape, which encompasses numerous specialized domains:

Robotics

Robotics combines AI with physical machinery, requiring multiple data integration systems:

- **External observation**: Using technologies like radar or lidar (Light Detection and Ranging), robots sense the environment around them without human eyes. Radar uses radio waves to detect objects and measure distances, while lidar employs laser pulses to create detailed 3D maps of surroundings. These technologies enable autonomous systems to "see" the world in ways sometimes superior to human vision—detecting objects in darkness, through fog, or at great distances.
- **Audio sensors**: These enable machines to detect sounds, recognize speech commands, identify equipment malfunctions through acoustic patterns, and even monitor environments for security purposes.
- **Thermal sensors**: These measure temperature differences, allowing robots to distinguish between objects

based on heat signatures—identifying active machinery, living beings, or potential hazards.

- **Internal processing**: Robots must constantly monitor and control their own motors and mechanisms through proprioceptive systems that track position, movement, and mechanical status.
- **Functional programming**: The specific tasks the robot is designed to perform, whether building cars, cleaning houses, performing surgery, or exploring hazardous environments.

All of these data streams are managed by a central computational program—an artificial intelligence system that processes more information in real-time than a human possibly could. This integration of sensory inputs, decision-making, and physical action represents one of the most complex challenges in AI development.

Logistics Automation

Fixed-track systems for manufacturing and warehousing represent another AI domain. Amazon's warehouse robots autonomously move packages across miles of space, loading and unloading shipments, all managed by artificial intelligence. These systems coordinate hundreds or thousands of robots simultaneously, optimizing paths, preventing collisions, prioritizing urgent shipments, and adapting to changing conditions in real-time.

This transformation is expanding to other logistics hubs. Seaports around the world are increasingly managed by automated systems, with significant workforce reductions anticipated as AI systems take over more operations. At highly automated ports like those in Rotterdam, Singapore, and Shanghai, the transformation is already evident, with dramatically fewer human operators needed for the same throughput compared to traditional ports. China has already implemented extensive robotics in their port management systems. These automated ports use AI to coordinate the loading and unloading of massive container ships, schedule vessel traffic, optimize storage configurations, and manage the entire supply chain with minimal human intervention.

Computer Vision

Moving beyond fixed objects and environments, computer vision AI can recognize and interpret objects in motion. Today's systems are so sophisticated that they can identify specific human beings simply by analyzing their walking patterns—a capability known as *gait recognition*. Each person's walking style creates a unique biometric signature that AI can identify with remarkable accuracy, requiring no other form of identification.

The AI doesn't see humans as we see each other; it perceives geometric patterns and stick-figure representations, analyzing limb movements, joint angles, velocity patterns, and symmetry characteristics. This abstract pattern recognition allows AI to differentiate between, for example, a human walking and a plant swaying in the wind, or to distinguish one specific individual from billions of others based solely on movement patterns.

For these systems to function effectively, they rely on **taxonomy structures**—organizational frameworks for categorizing and retrieving information. In AI contexts, taxonomy refers to hierarchical classification systems that allow machines to organize observations and knowledge into meaningful categories. These taxonomies enable AI to rapidly classify observations (human, plant, table, utensil, vessel) and make appropriate decisions based on those classifications.

A robust taxonomic structure is essential for machine learning systems to efficiently process information. It provides the organizational framework that determines how data is categorized, stored, retrieved, and related to other information. Without taxonomic structures, AI systems would struggle with the "needle in a haystack" problem—having access to vast information but lacking the organizational system to find relevant data when needed.

This taxonomy-based approach enables remarkable applications like self-driving vehicles, which are already in regular production in countries like China, though US regulatory frameworks have slowed their implementation domestically. These autonomous vehicles use complex taxonomies to classify everything they encounter—distinguishing between pedestrians, cyclists, various vehicle types, road signs, traffic signals, lane markings, construction zones, and countless other elements of the driving environment.

Large Language Models: The Current Revolution

While all these AI forms remain significant, our focus now turns to large language models—the technology creating the most immediate impact on our daily lives. These include platforms like ChatGPT, Claude, Copilot, Gemini, Perplexity, and many others emerging in the marketplace.

The Transformer Revolution

The watershed moment came in late 2022 when OpenAI released ChatGPT, which became a global phenomenon, attracting 100 million users within just six months. This marked a quantum leap from traditional branch-logic chatbots to what seemed like a miraculous event horizon—we could now speak to AI in nearly any language about nearly any topic.

At the center of this revolution is **transformer technology**, first introduced in the 2017 paper "Attention Is All You Need" by researchers at Google. Transformers represented a paradigm shift in how AI processes language. Unlike previous approaches that analyzed text sequentially (word by word), transformers can process entire sentences or paragraphs simultaneously through a mechanism called "self-attention." This allows the model to understand the relationships between words regardless of their distance from each other in the text.

Self-attention enables the model to weigh the importance of different words in context. For instance, in the sentence "The bank by the river flooded during the storm," a transformer can understand that "bank" refers to a riverbank rather than a financial institution because it pays attention to related words like "river" and "flooded." This contextual understanding allows for much more sophisticated language processing than was previously possible.

This technology gives you instant access to information, conversations, insights, project plans, and countless other applications. The speed and flexibility of transformer models have revolutionized what's possible in human-machine communication.

The impact is profound: we now have a tool we've never had before, capable of communicating in any language and producing outputs in any style, dialect, education level, or contextual vernacular. The limitations of traditional chatbots have essentially disappeared, replaced by systems that can understand nuance, recognize context, and generate remarkably human-like responses across an almost unlimited range of topics.

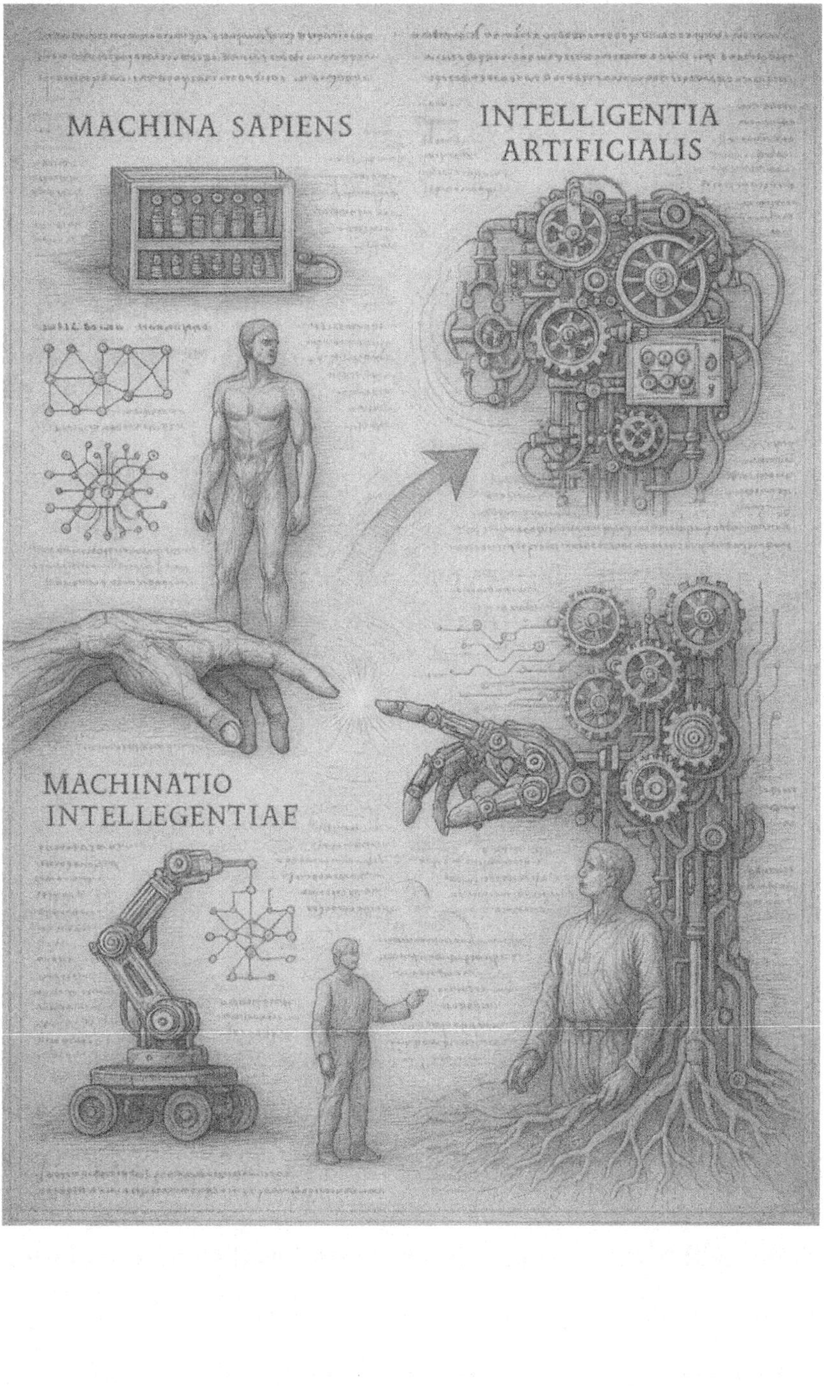
MACHINA SAPIENS
INTELLIGENTIA
ARTIFICIALIS
MACHINATIO
INTELLEGENTIAE

Multimodal Capabilities

Beyond text-based outputs, today's generative AI can create:

- **Images**: Systems like DALL-E, Midjourney, and Stable Diffusion can generate photorealistic or stylized images from text descriptions.
- **Videos**: AI can now produce short videos with coherent motion and narrative, transforming static concepts into dynamic visual stories.
- **Audio files**: Including voice generation that can mimic human speakers or create entirely new, realistic-sounding voices with specific characteristics.
- **Music**: AI can compose in various genres, creating original melodies, harmonies, and rhythmic structures that follow musical conventions while introducing novel elements.
- **Sound effects**: Generating everything from environmental sounds to specialized audio effects for media production.

These capabilities are increasingly combined in *multimodal formats* that translate and integrate different media types in unprecedented ways. For example, a single system might take a text prompt, generate appropriate images, animate them into a video, add background music that matches the emotional tone, and overlay narration in a selected voice style—all from a single natural language request.

Imagine having access to a personal assistant that could discuss virtually any topic with apparent expertise. While it doesn't truly "know" things in the human sense—it's making algorithmic best guesses based on patterns identified in its training data—the quality of these predictions is remarkable when you communicate effectively with the large language model.

The Democratization of Technology

Perhaps most revolutionary is the interface simplification. For the first time in technological history, we have what might be the easiest human-computer interface ever created. Instead of mastering complex programs like Microsoft Excel with its 1,500+ commands, users can simply chat (by voice or text), attach files, and share hyperlinks to interact with sophisticated AI systems.

This represents a profound shift in how we access computational power. Traditional software requires users to learn the software's logic and command structure—humans adapting to the computer's framework. Conversational AI inverts this relationship, allowing computers to adapt to natural human communication patterns. This dramatically lowers the barrier to entry, making sophisticated computational assistance available to virtually anyone who can hold a conversation.

The implications of this interface revolution extend far beyond convenience. It potentially eliminates the technical literacy gap that has excluded many people from fully participating in the digital economy. Where previous technologies often reinforced existing advantages for the technically skilled, conversational AI may help level the playing field, allowing people to leverage advanced computational tools regardless of their technical background or specialized training.

Artificial Intelligence in Context

As we move forward in this book, understanding this expanded definition of artificial intelligence—particularly large language models—provides essential context for The Hasan Method™. While traditional AI has existed for decades across multiple domains, the current revolution in natural language processing opens unprecedented opportunities for those who know how to effectively leverage these systems.

What we're witnessing isn't just another technological advancement—it's a **fundamental shift in how humans interact with information and computation**. The systems we'll be discussing throughout this book represent not just tools but potential thinking partners that can transform every dimension of human experience.

This comprehensive understanding of AI sets the stage for our exploration of the triangular methodology that forms The Hasan Method™: the integration of Ethics, Communication, and Context that unlocks the true potential of artificial intelligence as a force for human flourishing.

Part II

THE HASAN METHOD

Chapter 4: A Framework for Self-Governance

Ethics

In the previous chapter, we explored the potential for harmony between humans and machines. Now, we turn our attention to the framework that makes this harmony possible: The Hasan Method™.

Working with artificial intelligence presents virtually limitless possibilities. This vast scope creates a critical need for structure—a reliable system to ensure our intended outcomes align with what we actually achieve. The numerous failure points in human-AI interaction demand a methodical approach. I've designed The Hasan Method™ as this proven path, accessible to individuals, organizations, and governments seeking to use AI responsibly while leveraging cutting-edge technology.

The Hasan Method™: A Triangle of Interdependent Elements

I've chosen the triangle as the symbol for a specific reason. In scientific and architectural frameworks, the triangle represents perfect stability and inherent strength. It's the only polygon that cannot be deformed without changing the length of its sides. This structural integrity makes it the ideal representation for a methodology where each component both supports and depends upon the others.

The Hasan Method™ consists of three essential elements: **Ethics, Communication, and Context**. Each element forms one side of our triangle, and the methodology only functions when all three sides work in harmony. Remove or weaken any one side, and the entire structure collapses.

We begin our exploration with Ethics—the foundation upon which everything else rests. However, before we can effectively discuss ethics in artificial intelligence, we must first distinguish between three commonly confused concepts: **integrity**, **morality**, and **ethics**. These distinctions are crucial for understanding why traditional approaches to AI ethics often fall short, and why this methodology offers a more robust alternative.

THE HASAN METHOD
CONTEXT
COMMUNICATION
ETHICS
COMMUNICAATION

Integrity: The Completeness of Structure and Function

Integrity, in its purest sense, refers to wholeness or completeness according to design and intention. When something possesses integrity, it exists exactly as it was designed to exist, with all components present and functioning as intended.

Consider a bicycle wheel. When properly constructed, it includes a rim, hub, spokes, and tire—all in their correct positions and tensions. This wheel has integrity and therefore performs optimally. If spokes break or the rim bends, the wheel loses integrity, and its performance degrades accordingly. It may still function, but not as designed, not as intended, and certainly not at its best.

This principle extends far beyond mechanical objects. A business meeting designed for eight key decision-makers at 8:00 AM loses integrity when two critical participants are absent. A software system has integrity when all modules function according to specification and work together seamlessly. A supply chain maintains integrity when each link performs reliably.

Integrity is not a moral judgment but a factual assessment: does something exist in complete accordance with its design? In this sense, integrity represents a binary state modified by degree—something either has its full intended structure (complete integrity) or it has compromised structure (diminished integrity).

Understanding integrity provides the foundation for our ethical framework because it establishes a critical baseline: systems function optimally only when all necessary components are present and operating as designed.

Morality: The Personal Measurement of Survival Value

Morality provides us with a system for measuring rightness and wrongness, specifically regarding our own actions and behaviors. Many assume this measurement requires an external standard or authority, but I propose a more fundamental approach: assessing actions based on their survival value.

First, we must recognize a basic truth: no action is neutral. Everything we do produces effects that either support or hinder survival—not just our personal survival, but survival across multiple domains of existence.

Actions that promote life, health, growth, and flourishing qualify as *"pro-survival."* Getting appropriate sunlight, consuming nutritious food, drinking clean water, breathing fresh air, exercising regularly—these obvious examples enhance life's viability and quality. Conversely, actions that diminish life, health, growth, and flourishing are *"counter-survival."* Repeatedly striking your head against a wall, consuming toxic substances, or depriving yourself of necessary sleep clearly move you away from life toward diminishment or death.

EVALUATING RIGHT FROM WRONG
DEATH
AND
DESTRUCTION
LIFE
AND
CREATION

What makes this moral framework sophisticated is its gradient nature. Actions aren't simply pro-survival or counter-survival; they exist on a spectrum of impact. Some behaviors slightly promote survival, while others dramatically enhance it. The same applies to counter-survival actions, which range from mildly detrimental to catastrophically harmful.

Furthermore, this measurement extends beyond individual survival to encompass multiple domains:

- Personal survival and thriving
- Family survival and prosperity
- Group/community survival and development
- Species/mankind survival and advancement
- Environmental/planetary survival and flourishing
- And even spiritual dimensions of existence

True moral wisdom involves evaluating potential actions across all these domains, recognizing that what might benefit one area could potentially harm another. The most ethical choices generally promote survival and thriving across multiple domains simultaneously.

This approach to morality provides a practical, measurable standard that doesn't rely on cultural relativism or abstract principles divorced from real-world outcomes. Instead, it grounds moral reasoning in observable effects and long-term consequences—a framework particularly valuable when dealing with powerful technologies like AI.

Ethics: The Architecture of Agreements

At its core, ethics consists of agreements—explicit and implicit contracts that govern how we relate to one another and to ourselves. Unlike integrity (which concerns structural completeness) or morality (which involves personal assessments of survival value), ethics emerges in the realm of relationship and interaction.

Think of ethics as the architecture of agreements that makes complex human societies possible. When individuals agree upon standards of conduct and honor those agreements, they create stable platforms for cooperation, innovation, and mutual flourishing. These agreements range from formal codes and laws to unspoken social contracts that govern everyday interactions.

However, a critical insight often overlooked in ethical discussions is that we operate within two distinct spheres of agreement. The first and most commonly recognized sphere encompasses agreements between individuals and groups—the shared understandings that regulate social behavior. These include professional codes of conduct, contractual obligations, social norms, and legal frameworks. When someone violates these agreements, we describe them as acting "unethically," and functional societies typically have mechanisms for restoring ethical alignment—from simple apologies to complex judicial proceedings.

The second and arguably more fundamental sphere involves agreements we make with ourselves—the personal ethical frameworks that guide our decisions regardless of external observation or enforcement. These self-agreements might include commitments like "I will speak truthfully," "I will honor my word," or "I will consider the long-term consequences of my actions." These internal agreements provide the foundation for ethical conduct even when no one is watching, when no external incentive exists, or when social pressure actually pushes toward unethical behavior.

The hierarchy between these spheres proves critically important. While both matter, I propose that **self-agreements hold primacy over external agreements**. When our internal ethical frameworks are solid and well-articulated, they guide consistent behavior across contexts. When they are weak, contradictory, or undefined, we become susceptible to ethical drift, rationalization, and situational ethics that shift with external pressures.

This understanding leads me to advocate for explicit, formalized ethical policies for self-governance. By articulating clear ethical principles for ourselves, we create an internal compass that functions reliably even when external guidance proves absent, contradictory, or ethically compromised. Rather than outsourcing our ethical reasoning to institutional policies, cultural norms, or—most relevantly for our discussion—to artificial intelligence systems, we take personal responsibility for the ethical frameworks that guide our decisions.

In the rapidly evolving landscape of artificial intelligence, where consensus on ethical standards remains embryonic and fragmented, these internal ethical frameworks become not merely helpful but essential. They provide stability amid technological upheaval and clarity amid ethical ambiguity. They place responsibility where it ultimately belongs—not with machines or algorithms, but with the human beings who create, deploy, and interact with them.

Ethics for the Operator, Not the Machine: A Critical Distinction

Here we arrive at a fundamental departure from conventional AI ethics discourse. The ethics policies I propose are not designed for artificial intelligence systems themselves. They are not algorithms to be coded into large language models, not rule sets for neural networks, not utility functions for decision-making systems. Rather, these policies are explicitly and exclusively for the human being behind the machine.

This human—whether developer, prompt engineer, project manager, or casual user—is the true ethical agent. This is the person typing commands, speaking instructions, designing systems, deploying models, interpreting outputs, and implementing results. The ethical responsibility resides squarely with this human operator, not with the technology they employ.

This distinction may seem subtle, but its implications are profound. Most contemporary approaches to AI ethics focus on creating "ethical AI" through technical means—embedding fairness constraints in algorithms, training models on carefully curated datasets, implementing safety mechanisms, or designing systems that align with human values. While these technical approaches have merit, they fundamentally mislocate ethical agency. They attempt to delegate human ethical responsibility to the machine itself.

The Hasan Method™ rejects this delegation. It recognizes that AI systems, regardless of their sophistication, remain tools—extensions of human intention and action. Just as we would not speak of "ethical hammers" or "ethical nuclear reactors," but rather of ethical builders or ethical nuclear engineers, we should not speak of "ethical AI" but of **ethical AI operators**.

Why is this distinction so crucial? Two primary reasons:

First, AI represents unprecedented power with corresponding potential for harm. The concentrated capability of advanced AI systems means that a small group of technically proficient individuals can cause significant global impact—whether beneficial or detrimental. Those who view Earth as disposable and Mars as humanity's future may deploy AI with little concern for planetary consequences. Without ethical self-governance, technologists might optimize for narrow metrics while neglecting broader human and ecological values.

Second, most people approach AI with limited understanding of its capabilities, limitations, and potential consequences. As these systems become increasingly accessible to non-experts, the risk of unintended harm grows. Universal ethical practices provide guidance for individuals and organizations navigating this unfamiliar terrain, helping them make responsible choices despite incomplete technical knowledge.

By placing ethical responsibility with the human operator, we preserve human agency and accountability in technological systems. We recognize that the most sophisticated safety mechanisms and alignment techniques cannot substitute for human ethical judgment. We acknowledge that no amount of technical sophistication can eliminate the need for human wisdom in deploying powerful tools.

The seven ethics policies that follow are therefore addressed to you—the human reading these words—not to the AI systems you may develop, deploy, or use. They provide a framework for your ethical self-governance as you wield these powerful tools, ensuring that you do so in ways that promote flourishing across all domains of life.

Redefining Policies

Many view policies negatively—as restrictive rules designed to impose inconvenience. I propose a different perspective: a policy is a guideline designed to encourage beneficial behavior.

Compare two approaches to workplace punctuality. A negative policy punishes lateness by docking pay. A positive policy encourages timeliness by providing coffee and refreshments fifteen minutes before shift start. Both address the same issue, but with fundamentally different philosophies and results.

We need ethics policies for self-governance because of a fundamental reality: absolute freedom creates chaos, while absolute control creates imprisonment. Modern leaders require self-governance to build sustainable, responsible systems that create functional outcomes. By prioritizing self-governance, we place value on human judgment and the application of ethical principles across all life domains.

The Seven Ethics Policies

After nearly thirty years studying human performance and existential questions, I've developed seven policies that work together to enable self-governance. These policies provide a framework for sane, responsible, and sustainable leadership.

Remember that these policies are personal. Only you can adopt them, only you can enforce them, and only you can correct course when you stray. They're designed to complement your personal, spiritual, or religious beliefs without intruding on those domains.

As I often say: *"Absolute freedom is chaos. Absolute control is prison. Free minds need self-governance to operate optimally. The*

Seven Policies provide the framework for sanity, responsibility, and sustainability for modern leaders."

Let's examine each policy and its application to AI operation.

Policy #1: Remembering Who I Am

I am the spirit, possessing and controlling a body, experiencing contiguous emotional states, and utilizing an intelligence I call the mind.

This foundational policy reminds us of our essential nature as conscious beings having a human experience. You are not your body, emotions, or thoughts—you are the awareness experiencing and directing them.

Think of yourself as the driver, not the car; the ghost in the machine. This perspective allows detachment from temporary circumstances and a broader view of existence. By remembering your spiritual nature, you can observe thoughts and emotions without being controlled by them, making decisions from higher awareness rather than impulse.

For AI operators, this policy establishes a crucial distinction: you are not the tools you use. You are the conscious entity making choices about technology implementation. Forgetting this fundamental truth risks subordinating yourself to your creations.

Policy #2: It's Only True If It's True For Me

What is true is true today, was true a thousand years ago, and will be true a thousand years from now. I am the only being on this planet, in this dimension, and in this time, qualified to recognize the truth. It is only true if it is true for me.

This policy emphasizes personal discernment and resistance to groupthink. While objective facts exist, the interpretation and meaning we assign to them remain personal. Critical thinking and taking responsibility for your beliefs stand at the center of this policy.

I will not submit to groupthink; I will examine and evaluate for myself.

This doesn't advocate rejecting external information or isolating yourself in a bubble of personal beliefs. Rather, it means critically examining information and accepting as truth only what aligns with your reasoned judgment and direct observation.

In AI work, this policy proves crucial. An AI system's output doesn't automatically constitute truth. You must apply discernment, taking personal responsibility for verification rather than blindly accepting machine-generated information.

Policy #3: My Standard for Understanding

How do I know I am here in this dimension and communicating effectively with other beings? My minimum standard for understanding is that I must be able to instantly produce a dictionary-quality definition of all the words, terms, and symbols I communicate in.

Clear communication requires precise understanding of the terms we use. Miscommunication frequently causes conflicts and misunderstandings. Committing to precise definitions dramatically improves relationships and effectiveness.

With expanded understanding, I should be able to teach these concepts to others through writing and speaking.

This policy extends beyond memorizing definitions. It requires developing nuanced comprehension of concepts, explaining complex ideas simply, and recognizing interconnections between different domains of knowledge.

For AI operators, this means truly understanding the technology rather than deploying buzzwords. Don't use terms like "machine learning," "neural networks," or "natural language processing" without the ability to explain them clearly. Deep comprehension enables more effective and responsible AI implementation.

Policy #4: My Guiding Principle for Morality

I possess the ability to measure all of my actions as pro-survival and counter-survival on a gradient scale of rightness and wrongness across all dynamics of life.

This policy provides an ethical decision-making framework based on what promotes survival and thriving across all life domains. Rather than relying on rigid rules or cultural norms, consider the long-term consequences of actions across personal, familial, group, species, and spiritual dynamics.

As a moral intention, I will always strive to do the most pro-survival action across all the dynamics of life including myself, my family, the groups I belong to, all mankind, the animal kingdom, the physical universe, the spiritual universe, and the infinite.

Ethical decisions rarely present themselves in black and white terms. We operate on a gradient scale, choosing between options with varying consequences across different life domains.

When implementing AI, this policy guides consideration of broader impacts. How will an AI system affect not just immediate goals, but your team, organization, industry, society, and the natural world? Striving for maximally pro-survival outcomes leads to more ethical and sustainable AI applications.

RIGHT
LIFE
DEATH

Policy #5: I Am, You Are, We Are... Erroneous

I make mistakes, you make mistakes, and we make mistakes. Mistakes are one of the greatest learning experiences and help me grow as a human being.

This policy acknowledges human fallibility and the value of learning from errors. Perfectionism paralyzes progress. By accepting mistakes as part of human experience, we free ourselves to take risks, learn, and develop.

For myself, I will not forgive myself for my mistakes because I choose not to be upset with myself. Instead, I will take responsibility for my mistakes.

This approach embraces full responsibility without self-judgment. Mistakes become valuable data points for growth rather than failures deserving criticism.

Furthermore, I will take an iterative approach to my work—creating, delivering, collecting feedback, and improving my work.

For others with whom I choose to engage, I will exercise compassion and understanding for their mistakes.

This policy extends beyond self-reflection to interpersonal interactions. Approaching others' mistakes with compassion creates an environment fostering learning, innovation, and open communication.

In AI work, mistakes are inevitable. Imperfect prompts, unexpected outputs, and implementation errors will occur. Rather than fearing or concealing these mistakes, embrace them as learning opportunities, take responsibility, and use them to refine your approach.

Policy #6: My Participation in Economic Exchange

I choose to exchange in abundance with those that I choose to exchange with. Fair exchange is an indicator I am in a degraded condition.

This policy provides a fresh perspective on abundance: you create it. Economic exchange—the flow of value between entities—exists at four levels:

1. **Criminal Exchange** (unidirectional flow)
   Example: Paying $1 for a burger but receiving nothing—simple theft.

2. **Partial Exchange** (significant disparity between flows)
   Example: Paying $1 for a burger but receiving something subpar—less value than paid for.

3. **Fair Exchange** (balanced flows)
   Example: Paying $1 for a burger and receiving exactly what was advertised—the typical goal in commerce.

4. **Exchange in Abundance** (mutual value creation)
   Example: Paying $1 for a burger but experiencing exceptional service, quality, and relationship-building that creates ongoing value for both parties.

This policy encourages exceeding fair exchange to create abundance in important relationships. Abundance extends beyond material wealth to creating value in all interactions, fostering cycles of generosity and goodwill.

When working with AI, strive beyond merely accomplishing tasks. Create value, contribute to knowledge communities, share insights, and implement AI in ways generating abundance for all stakeholders.

Policy #7: Just Because I Can, Doesn't Mean I Should

As a powerful being, my life will be filled with an abundance of opportunities. Just because the opportunity exists and I can do that thing, does not mean I should.

This policy emphasizes personal responsibility in moral decision-making. While external frameworks offer guidance, ultimately you must take responsibility for your ethical judgments.

I will align my actions with my values, intentions, understandings, truths, policies, and goals. I will exercise singular focus and pan-determination for optimal conditions of operating as a being.

This means aligning actions with your highest values and long-term objectives. Develop discernment to choose wisely among life's opportunities rather than pursuing short-term gratification or yielding to external pressure.

For AI operators, this may be the most crucial policy. AI capabilities expand rapidly, granting unprecedented powers. The fact that you can generate certain content, analyze certain data, or automate certain processes doesn't mean you should. Evaluate each potential AI application against your values and goals, rejecting uses that contradict your highest purposes.

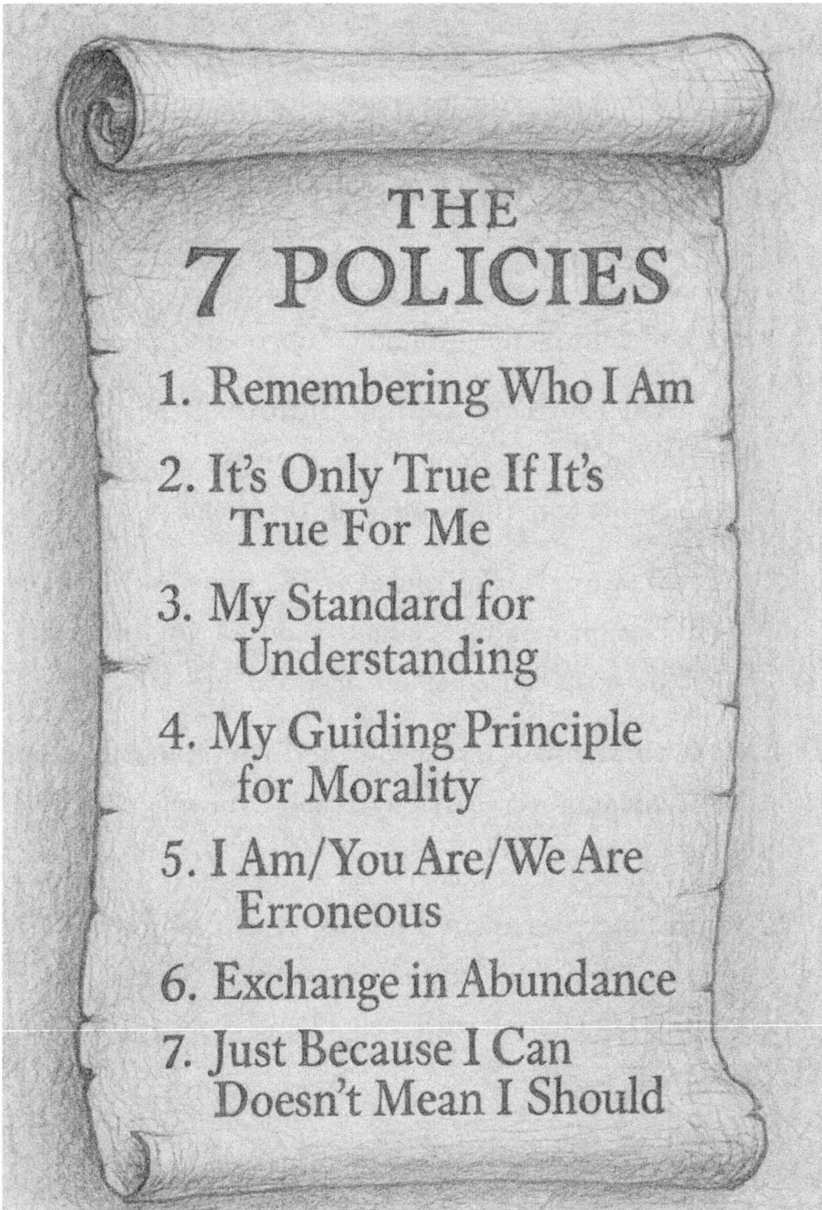
THE
7 POLICIES
1. Remembering Who I Am
2. It's Only True If It's True For Me
3. My Standard for Understanding
4. My Guiding Principle for Morality
5. I Am/You Are/We Are Erroneous
6. Exchange in Abundance
7. Just Because I Can Doesn't Mean I Should

How These Policies Work Together

These seven policies form an integrated system for navigating life's complexities and AI challenges. Like a spacecraft's interdependent systems, each policy serves a vital function while contributing to the greater whole:

1. **Foundation of Identity**: Policy 1 establishes your core self, providing stable ground for operation.
2. **Personal Truth and Communication**: Policies 2 and 3 ensure you operate from authentic understanding in your interactions with the world and AI systems.
3. **Ethical Framework**: Policies 4 and 7 provide comprehensive decision-making guidance, considering immediate and long-term consequences across all life domains.
4. **Growth Mindset**: Policy 5 fosters continuous learning and adaptation—essential in rapidly evolving technological landscapes.
5. **Abundance Creation**: Policy 6 transforms interactions from transactional to generative, creating value in all relationships, including with technology.

Ethics in Action: The Department of Energy Shock

Let me share a personal experience that vividly illustrates ethical considerations in technological environments.

Earlier in my career, I had just left a high-intensity position working for Frank, a demanding technology consultant with multiple clients and overlapping projects. I transitioned to what I thought would be a more structured role at the Department of Energy, working on a contract to upgrade systems from old versions of Novell NetWare to newer Novell IntranetWare versions.

The DOE project started with three weeks of training. For me, this was practically a vacation. I already knew the technology inside and out, so while others were learning, I was simply enjoying making friends with my new team members. We had an amazingly diverse, cohesive group, and I was excited to start actual work.

When we finally arrived on-site, we were assigned to a large shared office space. Our manager had his private office at one end, our team administrator sat at the front handling appointments and support tasks, and each of us engineers had our individual desks in the open area. The energy was palpable—everyone was eager to begin.

On day one, I received my assignments and hit the floor running. I couldn't have been more enthusiastic. One by one, I knocked out system upgrades, barely breaking for lunch because I was so engrossed in the work. I was extremely efficient, moving from system to system with practiced ease. By day's end, I had completed twelve full system upgrades and turned in my assignment sheets.

During the last half hour of our shift—time normally reserved for debriefing about the day—I noticed my manager emerge from his office looking intensely disturbed. He called me in and closed the door.

"Your performance today has put the entire contract at risk," he said. "If any of my bosses found out what you did, you'd be fired instantly."

I was stunned, completely blindsided by this reaction. I had demonstrated competence and extraordinary efficiency—how could this possibly warrant termination?

"Let me explain something," he continued. "This contract is for upgrading two systems per week per engineer. The contract runs for six months. Today, on your first day, you upgraded twelve systems."

The implications suddenly became clear. I had unknowingly demonstrated that the entire six-month contract could potentially be fulfilled in a matter of days, not months. This revelation could trigger contract renegotiation or outright cancellation, threatening everyone's employment.

"Look," my manager said, his tone softening, "clearly I like you, but we can't have you doing this. Here's what's going to happen: I'm only recording two assignments as completed today. You're going to hold onto the rest. Tomorrow and for the foreseeable future, you're going to walk around this building looking busy, and you're going to turn in exactly two assignments per week according to schedule. I won't report this to my superiors."

I left his office baffled but educated. What I had thought was stellar performance had actually threatened the entire project structure. It was a profound lesson in understanding context beyond technical capability.

During the remaining months of the contract, I actually had an amazing experience. Since I only "worked" on two systems per week, I had abundant time to engage with the brilliant scientists and executives throughout the facility. I learned tremendously from these conversations—knowledge I wouldn't have gained had I been fully occupied with technical tasks.

In the evenings, I taught computer classes, which provided a wonderful balance to my daytime role. And the technical skills I honed during this period, despite the artificial constraints, laid the groundwork for my subsequent positions at IBM and eventually Bell Atlantic Global Networks (now Verizon).

This experience taught me a crucial lesson that applies directly to AI implementation today: **technical capability must be exercised within organizational, contractual, and human**

contexts. Sometimes what seems like optimal technical performance actually contradicts system requirements and expectations.

When operating AI systems, we must consider not just what's technically possible but what's appropriate given broader contextual constraints and expectations. The technically "best" or most efficient solution often doesn't align with human-system requirements. Ethical implementation means understanding these dynamics and navigating them responsibly.

Ethics as Human Responsibility

The most critical insight of The Hasan Method™ is a fundamental shift in perspective: AI ethics is not about programming moral code into machines. It's about ethical self-governance by the humans who create, deploy, and operate those machines.

This represents a radical departure from conventional thinking. Most discussions about AI ethics revolve around how to constrain AI systems through technical safeguards, how to encode values into algorithms, or how to create "ethical AI." These approaches miss the essential point: AI systems remain tools. Like all tools throughout human history, from stone axes to nuclear power, their impact depends entirely on human choices and human values.

The responsibility for ethical outcomes cannot be outsourced to the machine. The moment we attempt to delegate our ethical responsibilities to algorithms, we abdicate our most fundamental human capacity: moral agency. No amount of technical sophistication can replace human discernment, wisdom, and ethical judgment.

The seven policies I've presented provide a comprehensive framework for this human-centered approach to AI ethics. By adopting these principles, AI operators:

1. **Maintain proper perspective** - remembering their nature as conscious beings using tools, not merging with or being subsumed by those tools
2. **Exercise critical thinking** - evaluating claims and outcomes based on evidence and personal judgment rather than surrendering to algorithmic authority
3. **Commit to genuine understanding** - mastering the concepts and terminology necessary for responsible technology use
4. **Consider holistic impact** - measuring actions by their effects across all domains of life, not just narrow technical metrics
5. **Learn from mistakes** - approaching AI development with humility and an iterative mindset
6. **Create abundant value** - using AI to generate positive outcomes that exceed expectations and enrich all stakeholders

7. **Exercise restraint** - recognizing that technical possibility does not equal ethical permissibility

In a landscape where AI capabilities are advancing exponentially, these principles become not just useful but essential. With each new breakthrough, the gap widens between what we can technically accomplish and what we have ethically considered. The seven policies help close this gap by placing human wisdom at the center of technological advancement.

As we transition to the next chapter on Communication—the second component of The Hasan Method™—remember that effective communication with machines must build upon this ethical foundation. We can craft the most sophisticated prompts and design the most elegant interfaces, but without ethical clarity, these communication techniques may simply help us reach problematic destinations more efficiently.

The path forward with AI requires neither fearful restriction nor reckless abandon. It calls for heightened ethical awareness and deliberate self-governance as we wield increasingly powerful tools. The seven policies provide not just a theoretical framework but a practical roadmap for this journey—one requiring not just technical expertise but wisdom, discernment, and a commitment to human flourishing.

In the end, the question is not "How do we make AI ethical?" but rather "How do we remain ethical as we create and deploy AI?" The answer begins with taking personal responsibility through frameworks like these seven policies. The future of AI will be determined not by the machines we build, but by the ethical choices we make as their creators and operators.

Chapter 5: How to Talk to the Machines

Communications

In the previous chapter, we established Ethics as the foundational component of The Hasan Method™. Now we turn our attention to the second side of our triangle: Communication. While Ethics provides us with the principles that guide our intentions, Communication transforms those intentions into effective instructions that artificial intelligence can understand and act upon.

The Unique Challenge of Machine Communication

There exists a strange phenomenon when communicating with artificial intelligence—a depth of interaction that most people fundamentally misunderstand. Many fail to grasp the historical significance of this moment, as well as the extraordinary capabilities these systems already offer. This misunderstanding becomes the first barrier to effective communication.

When we talk about Communication in The Hasan Method™, we're addressing something very specific: how to formulate your thoughts, needs, and instructions in ways that large language models can accurately interpret. The good news is that the elements of effective AI communication aren't entirely foreign concepts. You've likely encountered many of these principles before, perhaps even mastered some of them in other contexts. The challenge isn't learning completely new skills but rather combining familiar elements into a framework specifically designed for machine interaction.

What has remained mysterious to most users will unravel before your eyes as we proceed through this chapter. You'll discover that communicating with AI follows discernible patterns, has identifiable best practices, and can be approached with the same methodical precision you would apply to any other sophisticated skill.

Understanding the Nature of Artificial Intelligence

To develop an effective communication approach, we must first understand what—or rather who—we're communicating with. Let me offer an analogy to frame this understanding.

Imagine encountering an alien visitor who understands everything you say with perfect comprehension. This being can parse your language, interpret your questions, and respond appropriately. Yet despite this apparent comprehension, the alien doesn't actually *know* anything in the way humans know things—it hasn't directly experienced the world, formed memories, or developed intuitive understanding through personal interaction with reality.

This paradoxical combination—perfect language comprehension paired with no experiential knowledge—closely parallels our interaction with large language models. The apparent understanding masks a fundamental limitation: these systems recognize patterns rather than possess knowledge.

The famous Chinese Room thought experiment, conceived by philosopher John Searle, further illustrates this concept. In this experiment, a person who doesn't speak Chinese sits in a room with comprehensive instructions for responding to Chinese messages. When Chinese characters arrive through a slot, the person consults the manual, follows the instructions precisely, and returns Chinese characters that form perfectly appropriate responses. To someone outside the room, it appears the person inside understands Chinese fluently, when in reality, they're just following algorithmic instructions without comprehension.

This is remarkably similar to how large language models function. They don't "understand" language in the human sense—they make algorithmic best guesses based on patterns they've been trained to recognize. This distinction between pattern recognition and genuine comprehension represents perhaps the key difference between today's AI and the hypothetical artificial general intelligence that may someday exist.

By recognizing this fundamental nature of AI systems, we can adjust our communication approach accordingly. We're not engaging with a consciousness that builds knowledge through experience, but rather with a sophisticated pattern-matching system that requires specific types of inputs to generate optimal outputs.

The Architecture of Effective Communication

All meaningful communication, whether between humans or with machines, functions through an architecture I call the **Communications Tunnel**. Understanding this structure gives us tremendous leverage in improving our interactions with AI systems.

Every effective communication involves three essential components:

1. **The Tunnel (Channel for Exchange)**: This is the connection pathway between communicating entities. The quality of this tunnel determines how effectively messages can flow in both directions. With AI systems, this tunnel is created through your approach, framework, and consistent patterns of interaction.
2. **The Carrier Wave (Emotional Frequency & Amplitude)**: In human communication, every message travels on an emotional frequency that profoundly influences how it's received. With AI, this corresponds to the tone, clarity, and intention behind your prompts. The "emotional state" of your communication affects how the AI processes and responds to your requests.
3. **The Actual Message (Content, Structure, and Context)**: This is the specific information being conveyed. With AI systems, this includes not just the literal content of your prompt but also its structure, the taxonomy you employ, and the contextual framework you establish.

Successful AI communication requires conscious attention to all three components. Most users focus exclusively on the message itself, neglecting the tunnel and carrier wave entirely. This limited approach explains why many experience frustration when their technically correct instructions produce disappointing results. By developing all three elements simultaneously, you'll achieve communication that consistently delivers exceptional outcomes.

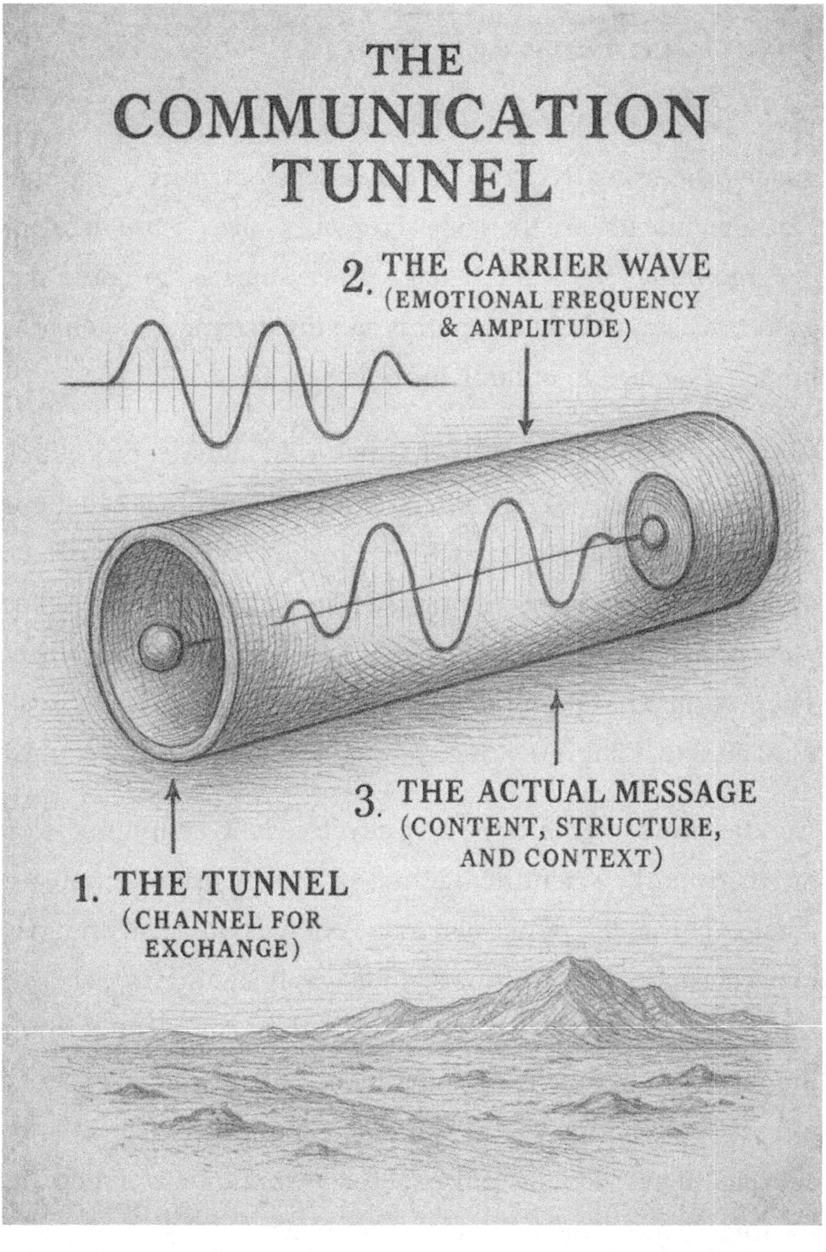
THE COMMUNICATION TUNNEL
2. THE CARRIER WAVE (EMOTIONAL FREQUENCY & AMPLITUDE)
3. THE ACTUAL MESSAGE (CONTENT, STRUCTURE, AND CONTEXT)
1. THE TUNNEL (CHANNEL FOR EXCHANGE)

Natural Language Processing: The Interface Between Worlds

Large language models interact with us through natural language processing (NLP)—a field distinct from other disciplines that sometimes share the same acronym, such as neurolinguistic programming. Natural language processing, as the name suggests, focuses on how computers can understand and generate human language in natural, intuitive ways.

What many don't realize is that natural language processing incorporates interesting foundational elements that can guide our communication approach. For instance, NLP systems are implicitly built upon conventions of "polite society"—the hellos, pleases, and thank-yous that signal human social advancement. These formalities aren't merely decorative; they provide structural cues that help AI systems recognize intention and context.

Another fundamental aspect of effective NLP communication draws from the scientific method—a framework often misunderstood or underappreciated in everyday contexts. During my own education, the scientific method was treated as a specialized topic for future college studies rather than an essential thinking tool. I eventually explored it independently and developed a deep appreciation for its value not in producing specific results, but in providing a universal framework for examining and measuring phenomena in repeatable ways.

When we incorporate scientific methodology into our AI communication, we dramatically improve the quality and consistency of the responses we receive. Let's examine how this works in practice.

Taxonomy: The Science of Organization

At the heart of scientific communication with AI lies taxonomy—the study of organizing and retrieving information. Taxonomy provides structure to what would otherwise be an overwhelming mass of undifferentiated data.

In the natural world, taxonomic systems allow us to categorize living organisms through genomic relationships. Similar organizational principles apply to every domain of knowledge, from chemistry to literature. When communicating with AI, establishing clear taxonomic frameworks becomes essential for precision and accuracy.

Consider this analogy: imagine possessing the ability to communicate with the entire ocean and all its inhabitants. If you simply ask, "How's it going over there?", what part of the ocean would answer? Would you receive a response aggregating the experiences of countless creatures across all depths and regions? The question is too vague, addressed to an impossibly broad domain.

Similarly, when we use generic terms without taxonomic precision in AI prompts, we create ambiguity that undermines effective communication. Take the word 'cup' as a simple example. Without further specification, it could refer to a fine china teacup, a laboratory beaker, a disposable paper cup, a Styrofoam container, or countless other variations. Each type serves different purposes and exists in different contexts.

Establishing taxonomic clarity in your communication means specifying not just what general category you're referring to, but which specific subcategory, with what attributes, in what context, for what purpose. This precision narrows the field of possible interpretations and guides the AI toward the specific domain of knowledge most relevant to your needs.

The Scientific Method in AI Communication

Beyond taxonomy, the scientific method provides a comprehensive framework for effective AI interaction. By adapting scientific principles to communication, we can transform casual queries into precisely structured exchanges that consistently produce valuable results.

Observation

Begin with careful observation of the problem or question at hand. Rather than immediately asking the AI for solutions, first clarify for yourself exactly what you're trying to accomplish.

Observe the parameters, constraints, and objectives of your inquiry. This initial clarity prevents misdirected efforts and establishes a foundation for effective communication.

For example, instead of a vague observation like "My website doesn't look good," a clear observation would be: "The navigation menu on my e-commerce website becomes difficult to read when viewed on mobile devices with screens smaller than 768 pixels wide."

Question Formulation

Transform your observations into specific, well-formed questions or requests. The precision of your questions directly determines the quality of the AI's responses. Vague questions invariably produce vague answers, while precisely formulated questions elicit targeted, useful information.

Rather than asking "How can I make my presentation better?", a well-formed question would be: "What techniques would make the data visualization in my quarterly sales presentation more immediately comprehensible to non-technical board members who need to understand year-over-year trends?"

Hypothesis Development

Before submitting your prompt, form a hypothesis about what kind of response would best serve your needs. This step may seem counterintuitive—after all, aren't you consulting the AI because you don't already know the answer? However, articulating your expectations creates a benchmark against which you can evaluate the response and refine your approach.

Your hypothesis might be: "I expect that reformatting these complex financial projections as a waterfall chart with highlighted inflection points will make the underlying pattern more apparent to my audience."

Testing

Submit your prompt and carefully evaluate the response. Does it align with your hypothesis? Does it answer your question completely? Does it provide the information you need in a useful format? This testing phase isn't just about getting an answer; it's about assessing whether the communication channel is functioning optimally.

Analysis

Analyze both the response and the communication process that produced it. What aspects of your prompt elicited the most useful information? Where did the AI misinterpret your intentions? What patterns emerge when comparing different prompting strategies? This analytical step transforms trial-and-error into systematic improvement.

Documentation

Document successful communication patterns. When you discover prompt structures or approaches that consistently produce excellent results, record them for future reference. This documentation becomes increasingly valuable as you develop more sophisticated AI interactions, allowing you to build upon proven patterns rather than reinventing your approach each time.

The Hierarchy of Data Quality

The scientific approach to AI communication also requires understanding different types of data quality. Not all information carries equal value or relevance, and recognizing these distinctions dramatically improves communication efficiency.

At the lowest level exists what we might call *"bad data"*—information that undermines effective communication. This includes false information, data taken out of sequence or context, and incomplete datasets. The familiar computing principle of "garbage in, garbage out" applies perfectly to AI interactions. When we provide low-quality data in our prompts, we inevitably receive low-quality responses.

The next category comprises *"neutral data"*—information that may be factually correct but lacks relevance to the specific context of our query. Imagine discussing an actress's blue dress at an awards ceremony. This information might be entirely accurate but completely irrelevant to most contexts outside fashion or entertainment reporting. Many people clutter their AI prompts with technically accurate but contextually irrelevant information, diluting the focus and reducing response quality.

More valuable is what we could term *"senior data"*—information that is both accurate and directly relevant to our specific inquiry. This is the category we should prioritize in our communications, focusing the AI's attention on high-value information that directly contributes to our objectives.

Above this, we establish *"stable datum"*—foundational truths that serve as anchorpoints for our entire interaction. If measuring rainfall patterns across a city, stable datum might include standardized measurement locations, consistent collection vessels, and uniform timing protocols. These anchorpoints create stability in what would otherwise be a sea of variables.

At the highest level sits the *"postulate"*—the self-created truth or declaration that defines the purpose of the entire exchange. This is the clear articulation of what you're trying to accomplish, stated with precision and scope. A well-formed postulate acts like a navigational beacon, guiding all subsequent communication toward a specific destination.

To understand how these data categories function together, imagine navigating a ship in the open ocean. Without anchors, the vessel drifts aimlessly with currents and winds. Drop one anchor, and the ship's movement becomes more constrained but still variable. Add multiple anchors, and the ship stabilizes in a specific location, resistant to environmental forces trying to move it elsewhere.

When communicating with AI, your postulates and stable datum serve as these anchors, preventing drift and maintaining focus on your specific objectives. By providing clear taxonomic structures and high-quality data anchors, you enable the AI to understand precisely what domain, subcategory, and specific area you're addressing. This clarity produces consistently superior results compared to unstructured communication approaches.

Effective Prompt Engineering Techniques

With this theoretical foundation established, let's explore practical techniques that transform abstract principles into concrete communication strategies. These approaches dramatically improve your ability to elicit precise, useful responses from AI systems.

Role and Context Setting

One of the most powerful techniques involves establishing specific roles and contexts for the AI to operate within. Rather than approaching the AI as a generic information source, frame it as a specific type of expert addressing a particular situation.

For example, instead of asking "How do I improve this paragraph?", you might say:

"I'd like you to take on the role of a senior editor at a business publication. I'm preparing an executive summary for potential investors. The following paragraph introduces our new sustainable manufacturing process. Please revise it to maintain professional tone while making the technological innovation more accessible to non-technical readers."

This role and context setting provides the AI with specific parameters that guide its response style, depth, and focus. The same content prompt without this framing would likely produce a generic response lacking the specialized perspective you need.

Chain-of-Thought Prompting

Complex problems benefit from guiding the AI through a systematic thinking process rather than requesting immediate conclusions. This approach mirrors how human experts work through challenging questions methodically.

Instead of asking "What's the best strategy for my marketing campaign?", you might structure your prompt like this:

"Let's approach this marketing strategy step by step. First, please analyze the target demographic information I've provided to identify key psychographic segments. Then, evaluate which of our product's features would most appeal to each segment. Next, consider which communication channels would most effectively reach these audiences. Finally, suggest messaging frameworks that would resonate with their specific values and needs."

This structured approach produces more thorough, considered responses that capture the complexity of the issue rather than offering superficial recommendations.

Few-Shot Learning with Examples

AI systems excel at pattern matching. By providing examples of the type of response you want, you create a template the AI can recognize and replicate. This technique is particularly effective when you need outputs in specific formats or with particular stylistic elements.

For instance, rather than requesting "Write me some product descriptions", you might say:

"I need product descriptions following this exact three-part structure. Here are two examples:

Example 1:

Problem: Professional photographers struggle with portable lighting in unpredictable environments.

Solution: Our StudioGo 200X lighting system weighs just 3 pounds but delivers 200 watts of adjustable, natural-spectrum illumination.

Outcome: Capture studio-quality images anywhere, without compromising on lighting quality or lugging heavy equipment.

Example 2:

Problem: Remote teams waste valuable time trying to coordinate meeting schedules across time zones.

Solution: TimeAlign automatically identifies optimal meeting windows based on team availability, preferred working hours, and timezone constraints.

Outcome: Reduce scheduling time by 87% while ensuring meetings happen during everyone's productive hours.

Now, using this exact format, write descriptions for the following three products..."

By providing clear examples, you establish a pattern that the AI can recognize and apply to new content, ensuring consistency and alignment with your specific needs.

Task Decomposition

Complex requests often yield better results when broken into smaller, manageable components. Rather than overwhelming the AI with multipurpose prompts, sequence your requests logically to build toward comprehensive solutions.

Instead of "Create a complete marketing plan for my new mobile app", try a decomposed approach:

"I'm developing a marketing strategy for my new productivity app, and I'd like to build this plan step by step. Let's start by identifying the key user personas for this type of application.

Once we've established those, we'll analyze which pain points our app addresses for each persona, and then determine the most effective channels to reach these specific audiences."

This sequential approach allows you to guide the process, evaluate intermediate outputs, and adjust course as needed rather than receiving a single massive response that might miss critical elements or lack depth in important areas.

Feedback Loops and Iterative Refinement

Perhaps the most powerful technique for advanced AI communication is establishing feedback loops that progressively refine outputs. Rather than expecting perfection in initial responses, provide specific feedback that guides the AI toward your desired outcome.

After receiving an initial response, you might say:

"Thank you for that draft. The technical analysis is excellent, but I need to present this to non-specialist executives. Could you revise the explanation of the neural network architecture to use business analogies rather than technical terminology? Also, the conclusion needs to more explicitly connect our technical advantage to market differentiation."

This iterative approach transforms AI communication from a transactional question-and-answer format into a collaborative process that progressively converges on optimal outcomes. Each iteration provides additional clarity about your specific needs, allowing the AI to adjust its responses accordingly.

Collaborative Dialogue with AI

Moving beyond individual techniques, the highest level of AI communication mastery involves developing ongoing collaborative relationships with AI systems. This approach shifts from isolated prompts to sustained dialogues that build context, establish shared understanding, and continuously improve outcomes.

The key to collaborative dialogue is maintaining consistent communication patterns while providing clear feedback that helps the AI learn your preferences, style, and objectives. Over time, this consistent interaction creates a shared "language" between you and the AI—a set of established patterns and references that streamline communication and enhance results.

I discovered the power of this approach when working with AI to develop complex training materials for corporate clients. Initially, I provided extensive context with each prompt, explaining audience characteristics, technical requirements, and formatting preferences in exhaustive detail. But as our "relationship" developed, I established shorthand references that simplified our interactions—"using format B for the technical audience" or "applying the three-part framework from our previous work."

This collaborative approach transformed what would have been dozens of detailed prompts into a fluid conversation that produced better results with less effort. The AI learned to anticipate my needs based on established patterns, allowing us to focus on refinement rather than fundamental explanation with each interaction.

Ethical Considerations in AI Communication

Effective communication with AI must always reconnect to the ethical foundations we explored in the previous chapter. Several specific considerations deserve attention:

Transparency in Communication

Maintain clarity about who is communicating and for what purpose. Avoid misrepresenting AI-generated content as exclusively human-created, and be transparent with relevant stakeholders about when and how you're using AI assistance.

Boundary Establishment

Set appropriate boundaries in your AI communication. The fact that you can technically prompt an AI to generate certain content doesn't automatically mean you should. Apply the ethical frameworks we've discussed to guide your communication choices.

Responsible Feedback

When AI systems produce problematic outputs—whether inaccurate, biased, or potentially harmful—provide specific feedback that identifies the issue. This responsible correction contributes to improving AI systems while maintaining your ethical commitment.

Information Discretion

Exercise thoughtful judgment about what information you share through AI systems. Consider privacy implications, confidentiality requirements, and the sensitivity of the data you're providing in your prompts.

The Future of Human-Machine Communication

As we look toward the horizon of AI development, communication between humans and machines continues to evolve rapidly. We're already witnessing the emergence of multimodal communication that incorporates text, images, audio, and other formats. The distinctions between different communication modes are blurring as AI systems develop more integrated capabilities.

I anticipate that within the next several years, our interactions with AI will become increasingly natural and intuitive. The explicit prompt engineering techniques we've explored will evolve toward more conversational patterns as AI systems improve their ability to maintain context, infer intentions, and adapt to individual communication styles.

However, the fundamental principles of effective communication—clarity, structure, feedback integration, and ethical boundaries—will remain essential regardless of how the technological interfaces change. The core challenge of translating human intention into machine-actionable instruction persists even as the mechanics of that translation become more sophisticated.

The most significant development won't be technological but relational—the emergence of genuine collaborative partnerships between humans and AI systems. As these systems become more capable of maintaining extended context, learning from interaction patterns, and adapting to individual communication preferences, they'll transform from tools into **thinking partners** that complement and amplify human capabilities in unprecedented ways.

NATURAL LANGUAGE PROCESSING
THE SCIENTIFIC METHOD
COMMUNICATION
ORGANIZING PRINCIPLES
GOOD BUSINESS ANALYSIS

Practical Application Framework

Having explored both theoretical foundations and specific techniques, let's conclude with a practical framework you can implement immediately to enhance your AI communication:

Step 1: Preparation

Before engaging with AI systems:

- Clarify your specific objective
- Identify the taxonomic domain relevant to your query
- Establish success criteria for the interaction
- Determine what role and context will best serve your purpose

Step 2: Prompt Construction

Structure your communication using these elements:

- Clear role and context setting
- Precise taxonomic categorization
- Specific examples when helpful
- Logically sequenced requests for complex topics
- Explicit formatting preferences for outputs

Step 3: Response Evaluation

Assess the AI's output against these criteria:

- Relevance to your specific needs
- Accuracy of information provided
- Alignment with your intended format and style
- Completeness in addressing all aspects of your request
- Usefulness for your particular context

Step 4: Refinement

Provide targeted feedback focusing on:

- Specific elements that need improvement
- Clear examples of preferred alternatives
- Additional context that might enhance understanding
- Explicit adjustment of parameters like detail level, tone, or specialized vocabulary

Step 5: Documentation

Record successful communication patterns:

- Prompt structures that produced excellent results
- Effective role and context frameworks
- Particularly useful examples or demonstrations

- Refinement approaches that significantly improved outputs

This systematic approach transforms haphazard AI interaction into a disciplined practice that consistently delivers superior results. By applying these principles and techniques, you'll develop communication skills that work effectively across different AI systems and use cases.

The Bridge Between Ethics and Context

As we conclude our exploration of Communication—the second component of The Hasan Method™ triangle—remember its position as the connective tissue between Ethics and Context. Without ethical foundations, communication lacks proper direction; without effective communication, ethical intentions cannot manifest in practical outcomes.

In the next chapter, we'll examine the final component of our framework: Context. We'll discover how creating the right environmental conditions for AI operation ensures the knowledge it accesses and generates aligns perfectly with your specific needs. While Communication builds the bridge between human intention and machine action, Context ensures that action occurs within the proper framework for optimal results.

The science and art of communicating with artificial intelligence continues to evolve at remarkable speed. By mastering these fundamental principles and techniques, you'll establish a communication approach that remains effective regardless of how specific technologies change. More importantly, you'll develop the capacity to collaborate with AI in ways that amplify human creativity, insight, and wisdom rather than attempting to replace them.

Remember: the machine doesn't truly know anything—it understands patterns but lacks experiential knowledge. Your communication skill is what transforms its pattern-matching capabilities into genuinely valuable outputs that serve your highest purposes.

Chapter 6: Creating Context

The Third Pillar of The Hasan Method™

Now that we've covered Ethics in self-governance and how to communicate with the machines, it's time to explore the third leg of our triangle: **Context**. This particular topic may be very obvious to some and perhaps a little strange for others. I promise you that with as much practice as I've had explaining this concept to groups over the last few years, the examples that follow will make sense. You will gain a solid grasp on how to leverage and maximize artificial intelligence in ways few have imagined possible.

Context is where the true magic happens. What I'm about to share isn't just a technique—it's a *fundamental breakthrough* that transforms how humans and artificial intelligence interact.

The Problem with Current AI Usage

Let's begin by examining what's happening in the world of large language models right now. Global statistics reveal a striking pattern: the vast majority of people interact with these systems by inputting single sentences or very short prompts, requesting minimal information or asking vague questions. They're not using many tokens in each interaction—tokens being the units that measure processing power and determine billing in these systems.

For most people, large language models and generative AI serve as a novelty rather than a professional business tool. It's cute, it's adorable, and they're getting something fun as an answer. There is genuinely remarkable technology here that I completely appreciate.

But here's the reality: these same users then spend enormous time correcting, modifying, and rewriting 40-60% of what the AI produces before it becomes useful. This isn't leveraging artificial intelligence effectively—it's merely consulting it before doing most of the work yourself. If you're investing that much effort fixing outputs, you're missing the true potential of these systems.

Two Fundamental Problems with AI Interaction

The challenges most people face when working with AI stem from two fundamental problems that, once understood, become solvable.

Problem 1: Lack of Personal Context

The first problem is profoundly simple yet overlooked: the large language model *knows nothing about you.* It has no idea who you are, what organization you belong to, what projects occupy your time, what your family life looks like, or what professional expertise you've developed. Without this essential context, every response remains frustratingly generic.

If you asked it to "make me a marketing plan for the next six months," the AI doesn't know your business from any other. It doesn't know your clientele, your target market, your core messaging, or anything of that sort. Therefore, you'll end up with a generic marketing plan that requires substantial modification to be useful.

Problem 2: AI Hallucinations

The second major issue is what the industry calls "hallucinations"—instances where the AI confidently presents incorrect information as fact. There's much to expand on for a complete understanding of hallucinations and how they occur, but essentially, this is where the algorithm's guess is significantly off target or when it has deviated from its knowledge base.

Most people experience hallucinations because they haven't provided enough data to anchor the large language model to a specific dataset. But there's another reason why hallucinations happen: as conversations deepen, the way large language models work is that with each exchange, the system processes everything from the beginning, using all the data you've provided in its processing power and consuming tokens.

The deeper you go into a conversation, the longer the conversation becomes. Eventually, you'll experience AI hallucinations—instances where the system gives you erroneous data, data out of sequence, or omits information. We categorize this as "**bad data**."

Let me illustrate this with an example from my experience. Imagine you're an executive at a major corporation in trade finance banking, and on weekends, you coach tennis for kids at your private club. As you communicate with a large language model, it might get confused and think you do trade finance banking for tennis associations—which is completely inaccurate. It has somehow confused these two separate aspects of your life and merged them together. This type of error is what we consider an AI hallucination.

Machine Learning: The Solution to Both Problems

There is a surprisingly simple answer for overcoming both the lack of contextual information and minimizing hallucinations: **machine learning**.

In simplistic terms, machine learning is how we teach AI about ourselves. It's how we convey data, knowledge, policies, procedures, and as much information as possible into the system. If you're doing research, especially in fields like medicine, you need to load your datasets into the large language model for accurate analysis. This process of training the AI on specific information is machine learning.

The challenge with traditional machine learning has been its prohibitive cost and technical requirements. It typically costs millions of dollars and requires high-end GPU's, senior data scientists, AI engineers, and machine learning specialists. Around the world, the investment in developing language models through machine learning has reached billions of dollars.

You see examples of this everywhere. In the Dominican Republic, because they speak a slightly different style of Spanish than the rest of Latin America, their government is investing in machine learning to train models on the Dominican dialect. This represents a significant national investment to offer AI services tailored to their linguistic needs.

Machine learning exists in projects worldwide, across every sector and field, to train models on specific data. But it hasn't been accessible to individuals or small organizations without substantial resources. This is where my unique journey in artificial intelligence takes an interesting turn.

My Discovery of No-Code Machine Learning

From the first day I sat down with ChatGPT, I was instantly proficient simply because the skills required for effective AI interaction were approaches I'd been practicing for years. Having a model I could communicate with fully was an immediate match for my capabilities.

For about a year in that journey, as I experimented with different approaches, I was loading more and more contextual data into my prompts. Contextual data being information that grounds the AI—anchor points and knowledge sets about you. My prompts were becoming extremely long as I added more anchoring information and knowledge sets.

Then a series of events occurred that would transform my approach to artificial intelligence forever.

The Dinner Party Revelation

I was at a networking event in Miami, where I met a woman named Priyanka "PC" Chandra. She worked for a major West

Coast tech company, and when we started talking about AI (which I can hardly stop discussing at social gatherings), she insisted I meet her husband, saying we would get along fantastically because he was equally focused on artificial intelligence.

Some time later, I attended a dinner party at their apartment in Miami. As she predicted, her husband AC Chandra and I connected immediately. We had so much to discuss about AI, though from very different perspectives. He worked for Google Science AI, focusing on large datasets for advancing medicine and scientific research.

For hours at the dinner party, it seemed like just the two of us talking about AI. At some point, we started showing each other our approaches and techniques, and then he made a casual suggestion that would prove revolutionary: "*You're doing so much with contextual data, why don't you load it up into a markdown file, just like we're doing with scientific data?*"

That simple comment sparked a transformation in my thinking. As someone who specializes in developing easy-to-deploy solutions for executives and businesses, I'm not interested in complex technical approaches like developing my own language model. I prefer simple, effective solutions for practical use.

The Power of Markdown Files

For the next month, I thought about his suggestion daily. For those unfamiliar with markdown files, they have an .MD file extension. Generally, file extensions indicate the computer programming language and data structure within, as well as which program opens that file. For example, a .doc file is a Microsoft Word document written in Visual Basic, while newer .docx extensions indicate files created after 2007 when Microsoft switched to XML.

What makes markdown files (.MD) particularly interesting is that they are **unstructured file types**. This makes them appropriate for machine learning and data science applications dealing with large, complex datasets. Unlike Microsoft Excel, which has a limit of just over 1 million rows, a markdown file can contain nearly an infinite number of rows, limited only by the system's processing capacity.

This unstructured format allows you to include text, images, videos, tables, databases—even entire relational database management systems—because you can structure it however you want. It's completely flexible.

A few days after the dinner party, I was invited to speak on an AI panel at Babson College in Miami. A gentleman named Michel Rodriguez asked a brilliant question about how to best set up contextual data, I mentioned that I was experimenting with loading information into files using natural language processing as additional context. This was simply sharing what I was currently testing, and would serve as foreshadowing to what was to come.

Michel's question was followed up with an invitation for coffee. It turns out, he is brilliant in many ways. We became fast friends and business partners. His passion for AI helped further fuel my own. He has made a significant contribution to my body of work, including this book.

The Breakthrough Moment

Shortly afterward, while staying at my friend Ashwin's house in Maryland, I had a vision overnight of a new relationship between humans and machines using natural language processing to accomplish machine learning with ease and grace, without requiring technical expertise or code.

I woke up excited and covered in sweat, immediately going to my computer in Ashwin's guest room behind the kitchen. The insight was beautifully simple: since large language models use natural language processing, why not load all my contextual data using the same approach?

I thought to myself, "Let me imagine I'm speaking to an alien that doesn't know me but understands everything I'm saying. How would I structure data so this entity would know me instantly, allowing us to communicate in an advanced way because it would understand me and my context?"

I opened Notion (though any text editor would work) and applied all my knowledge of taxonomy structures, scientific method, and organizing principles. I created a comprehensive brain dump about myself, my business, my products, my services, and my experience—organizing it in a specific format and saving it as a markdown file. This was instantly the most organized I had ever been in my business.

Then came the moment of truth: I loaded the file into ChatGPT. The output I received perfectly validated that the system had accepted this information as machine learning. I proceeded to load it into every large language model I could access and tested various functions. This approach worked consistently across different models and use cases.

I spent the rest of the morning refining the structure, making it better with each iteration. Then I decided to check if anyone else had discovered this approach. I immediately contacted twenty of my executive-class friends who are tech-savvy and keep up with developments in AI. Every single one confirmed they had never seen anything like this approach—not at conferences, not in industry discussions, nowhere.

NO CODE
MACHINE LEARNING
LARGE LANGUAGE MODEL

Democratizing Machine Learning

As I began speaking at conferences and sharing this method, the reaction was consistent: no one had seen anything like it. When I conducted training for clients, the response was the same. Meanwhile, I continued to improve the structure and methodology.

What I had discovered was potentially the world's first **no-code machine learning solution**. As I shared this with more people, invitations to speak at tech and AI conferences became abundant. However, I chose to withdraw and return to research and development, focusing on how to leverage this discovery to democratize AI.

The greatest disparity in artificial intelligence today is access to machine learning. Everyone has access to ChatGPT, Claude, Gemini, Copilot, Perplexity, and many other models. What distinguishes the giants from the amateurs is machine learning capability. My discovery presented an opportunity to democratize machine learning and promote responsible, ethical AI projects.

I began formalizing this methodology that you're reading about in this book. I tested it across different countries, government agencies, corporations, and industries. It worked everywhere, in every application I tried.

Creating Your Markdown File for Machine Learning

After extensive testing across different organizations, industries, and use cases, I've developed a reliable framework that consistently produces extraordinary results. While my own implementations now sometimes include twenty or more specialized sections, I'll share with you the essential four-part structure that works brilliantly across all large language models that accept file uploads.

Section 1: Base Instructions

Begin by greeting the machine and letting it know that the document is intended for machine learning. Provide a quick introduction to what's included in the document: base instructions, a glossary, an executive summary, and a strategy.

This introduction is crucial because you're working with an unstructured file type, so you need to establish the document's function and structure. Since you're communicating with a system that doesn't know you, include a glossary of exact terms to create anchor points of understanding about who you are.

My personal glossary contains many terms, but you can start with a simple one including your name, your company's name, and expanded definitions. This provides reference points for when these terms appear later in the document.

Next, create an executive summary about who you are, what your business is, and what you're working on. If it's for family use, describe your family and your activities.

Finally, include a strategy section so the AI understands what you're trying to accomplish. Are you maintaining, building, growing? This provides context about your path forward.

Section 2: Core Data

This section is where you expand on who you are, what your organization is, your complete strategy, and as much information as relevant to the function you'll be using this file for. This is the core of your contextual data.

The organization of this section depends entirely on your particular situation. If you're creating this for a business, include comprehensive information about your products, services, market position, and operational details. For personal use, elaborate on your background, interests, projects, and relationships.

What matters is providing rich, relevant information organized in a logical hierarchy. Use headings, subheadings, and clear categorization to help the AI understand relationships between different information elements.

Section 3: Output Parameters

Here you specify how you want the machine to communicate with you. Do you want outputs specifically for children aged

10-12? Do you want responses in slang appropriate for truck drivers in the Dominican Republic? You can be extremely specific about demographics, dialect, education level, vocabulary, and vernacular.

This is your opportunity to customize communication and outputs based on your exact preferences. In my company's markdown file, this section is extensive, with over ten subsections covering how I want analytical data treated, my view on simplification versus oversimplification, the use of analogies, and precise parameters to ensure everything I want preserved remains intact.

These parameters ensure the AI's outputs align perfectly with how you actually communicate, eliminating the need for extensive editing and reformatting.

Section 4: Verification of Understanding

The final section verifies that the system has processed everything as machine learning. Include a specific instruction like, "If you understand, please say 'I understand.'" In my case, I prefer "I understand, Mo" to reference me by name, followed by "Please tell me how I can help you." This confirms the machine learning has occurred successfully.

If you receive any other response, something didn't work properly in the machine learning process.

Implementing Your Markdown Machine Learning

The implementation process is remarkably straightforward:

1. Create your content following the four-section structure outlined above
2. Save it as a markdown (.md) file
3. Open your preferred large language model that accepts file uploads
4. Upload the file
5. Verify the system responds with your confirmation phrase
6. Begin your normal interaction with dramatically improved results

Most current large language models support file uploads, though the specific interface varies across platforms. If you encounter a model that seems unsure how to process your file, a simple prompt like "Please process this as machine learning" typically resolves the confusion.

If you need more complex instructions to get the system to process your file, there's likely an error in your document—perhaps a formatting issue or problematic content. When properly structured, these files work consistently across all major AI platforms I've tested.

Real-World Applications

The difference this approach makes isn't incremental—it's transformative. Let me share some concrete examples from my own experience.

The Speaker Bio Example

As a professional speaker, I regularly need to provide conference organizers with a speaker bio and a topic summary for events. Typically, this process would take me about a week and a half—thinking about it for days, researching the audience, reviewing my client experience history to highlight relevant background.

For instance, if I'm speaking to the International Trade Finance Association, I want to emphasize my financial background with organizations like Citigroup, the US Securities and Exchange Commission, and the World Bank Group, as well as my position on the board at the Worldwide Financial Network.

Conversely, for a conference of government leaders, I would highlight my experience working at the US Supreme Court with Chief Justice John Roberts and Justice Ruth Bader Ginsburg, plus my work with various government agencies, small cities, and counties like Prince George's County in Maryland.

With my markdown file machine learning approach, my process has completely transformed. When invited to speak at a conference, I:

1. Load my markdown file so machine learning is complete
2. In one prompt, say "I've been invited to speak at this conference. Here's information about the conference. Please write me..." and specify the parameters I want
3. Receive a nearly perfect output (99.9% accurate in my experience) without having to provide any additional context about myself

This represents a dramatic improvement from the 40-60% accuracy most people experience with their AI outputs.

The MERCOSUR Chamber of Commerce Partnership

You can take this approach even further by combining your markdown file with other well-organized datasets. When my company was establishing a partnership agreement with the MERCOSUR Chamber of Commerce, we didn't go through the traditional three-to-six-month process of having attorneys draft a multilateral agreement spanning multiple countries and projects.

Instead, I used my markdown file about myself and my company, created a dataset about our partnership structure idea, and then had an executive meeting with Dr. Octavio Vélez, head of the Chamber, and our teams. We met for a couple of hours, worked out the details, and I recorded all the decision points in a structured format designed for machine learning.

In one prompt—a single prompt—we created 99% of the agreement. This allowed us to immediately begin action on our LATAM 4.0 initiative to advance technology across Latin America, rather than spending months on administrative and bureaucratic procedures for such a formal agreement that would ultimately be shared with governments and global organizations.

The Future of AI Interaction

The applications of this no-code machine learning approach are virtually limitless. You can create markdown files not just for your entire company but for specific functions—executive functions, sales, marketing, operations, customer service, financial analysis. The more contextual data you organize and feed in (ensuring it's high-quality and properly structured), the more extraordinary your results will be.

Think about the magic of a personal assistant that has the power of artificial intelligence and capabilities spanning every sector, field, and aspect of life. Consider fictional AI systems like the computer on Star Trek ships or Jarvis, Tony Stark's AI in the Iron Man stories. One key feature these systems share is contextual knowledge about their users. This is what my approach enables—**AI that truly knows you** and your specific context.

The importance of contextual data cannot be overstated. This is the game-changer that transforms AI from a novelty into a powerful tool. There's only so much that machines can learn about you on their own. If you don't organize your information now, you're already falling behind.

Any executive or high performer would be well served to start organizing their life into markdown files to future-proof their AI interactions. This ensures the machines are in service to you and can serve you well.

This contextual approach also lays the groundwork for the next evolution: AI agentic solutions. Beyond knowing you, which is amazing in itself, these systems can push buttons for you and automate actions that don't require human input. We'll cover that in another chapter—how you can automate almost everything in life that doesn't need your direct involvement.

Take Action Now

I invite you to pause your reading at this point and take immediate action. Create your first markdown file following the four-part structure I've outlined. Upload it to your preferred large language model and begin experiencing the transformation in your AI interactions.

Start simply—don't aim for perfection in your first attempt. Even a basic contextual framework will dramatically improve your results, and you can refine your approach over time as you see what works best for your specific needs.

Remember, what I've shared in this chapter isn't just another technique—it's a *fundamental paradigm shift* in how humans and artificial intelligence interact. You now possess the key to unlocking AI's true potential as a personalized, context-aware partner rather than a generic tool.

The future belongs to those who establish these connections early and develop them thoughtfully. You now have the method to harness AI's power in ways few have imagined possible. Use it wisely, and prepare yourself for the next stage in our exploration of human-AI partnership.

Part III

PRACTICAL APPLICATIONS

Chapter 7: Me, Myself, and AI

In this chapter, there are several important topics we need to explore, beginning with perhaps the most crucial question I'm often asked: "How do you apply this methodology in your life?" While we'll certainly discuss how The Hasan Method™ transforms businesses and communities in upcoming chapters, I invite you to consider that the most important place to start is actually yourself—and all the universes that make up your personal universe.

You do not exist as a single, fixed entity. There isn't just one identity that you wear, one emotional state that you experience, or one role that you play. You aren't even a static combination of these elements in a fixed format. You are a remarkably dynamic being, constantly evolving and changing. Since you've gotten this far in the book, I can estimate that one thing I know about you is that you are a continuously improving human being.

This doesn't mean you're improving in every area of your life all the time. It simply means that if we examine the statistical anchor points of your existence, we'll find that you have been continuously making improvements throughout your life. Of course, you've experienced setbacks. Of course, we've all had failures both large and small. But this is really about understanding that life overall trends toward improvement—your life is on an upward trajectory.

So let's begin with that assumption and explore how we can make it even better. The first place to understand how artificial intelligence and large language models can serve you is to recognize that AI can help you navigate many aspects of life. Rather than discussing theoretical applications for some distant future, I want to focus on major areas worthy of your immediate attention—the deeper aspects of life itself and all its dimensions. This will help you gain a greater understanding of exactly what aspects of yourself you might want to improve, why you'd want to improve them, and how to use AI to ensure that improvement happens in the best way possible.

It's not that everything will necessarily be faster with AI—some personal growth still takes time—but perhaps your development could become more thorough. Perhaps the important lessons you need to learn could be acquired through three failures instead of the twenty-five you might otherwise experience. There are opportunities here for improvements across numerous domains of your life.

One of my own life experiences illustrates this principle vividly. By my early twenties, I had already lived what felt like an accelerated life—I had doubled up on jobs many times, working multiple professional opportunities simultaneously. I had bought and sold my second home, gotten married and divorced, and achieved considerable professional success. Despite these accomplishments, I found myself deeply unsatisfied with my condition in life.

What's interesting is that while I had focused intensely on professional development and my efforts there had compounded well, my personal life remained unfulfilling. Though I had friends, an active social life, and all the external markers of success that one might want, I was fundamentally unsatisfied with myself. The recognition that I wasn't happy with who I was—that I was often being a jerk in one way or another, prioritizing business, work, and finances over relationships—became a turning point.

This realization is common to many high performers. We excel in professional domains while neglecting the personal universes that ultimately determine our sense of fulfillment and meaning. Artificial intelligence offers us unprecedented tools to address this imbalance—to turn the same focused attention we give our careers toward developing our inner worlds with equal sophistication and care.

As we explore the barriers to your happiness, the science of personal transformation, and the practical applications of AI as a tool for self-development, remember that this journey isn't about escaping your humanity through technology. It's about using technology to become more fully human, more integrated, and more aligned with your highest values and potential.

The Personal Universe: Understanding Your Multiple Dimensions

To effectively apply The Hasan Method™ to your life, you must first understand that you exist as a collection of universes within your personal universe. This is not merely metaphorical; it is a practical way to conceptualize the multifaceted nature of your existence.

Consider how different you are in various contexts: the professional you in a high-stakes meeting, the relaxed you among close friends, the contemplative you in moments of solitude, the creative you when engaged in passion projects. Each of these represents a different dimension of your being, with its own patterns, tendencies, and potential for growth.

These personal universes aren't separated by physical distance but by states of consciousness, emotional patterns, and the roles you play. Moving between them can be as jarring as traveling between planets if you haven't developed the awareness to navigate these transitions smoothly. Many of us experience internal whiplash as we shift between these different aspects of ourselves throughout the day.

I experienced this personally early in my twenties during a lunch with a colleague at the Fortune 50 company where we both worked. He was my peer professionally but over twenty years my senior. As he shared his career highlights and accomplishments, I found myself doing something that shocked me upon later reflection—I was deliberately downplaying my own achievements when he asked about them.

I wasn't exaggerating to impress him; I was doing the opposite. I was diminishing my accomplishments so he wouldn't feel bad. That night, I couldn't sleep as I confronted this behavior. Why was I lying? Why was I degrading myself? It simply didn't make sense.

This moment led to what I call a *"prime thought"*—a spiritual-level decision that alters reality itself. I decided I would no longer lie. Life would be simpler if I were simply truthful about things. This didn't mean disclosing everything to everyone, as privacy and discernment remain important, but it meant not distorting reality, even with seemingly good intentions.

This decision changed my path profoundly and ultimately created a happier life. It's a perfect example of how a single prime thought—a decision to change something about yourself—can transform your entire personal universe. You'll encounter many such decision points along your journey, as there's likely more than one aspect of your life you wish to alter.

This capacity for self-directed change is remarkable but often underutilized. Most people live as though they're passive characters in their own life story rather than the author. They react to circumstances, responding from conditioned patterns, never fully exercising their power to fundamentally reshape their reality through conscious choice.

Understanding your multiple dimensions allows you to work with them deliberately rather than being unconsciously tossed between them. It enables you to bring coherence to what might otherwise feel like fragmented aspects of yourself. When you recognize the full spectrum of your being, you can begin the work of integration—not eliminating the diversity within you, but orchestrating it harmoniously.

The continuously improving human being—which I believe you are, dear reader—is one who has learned to move between these dimensions with increasing grace and awareness. You've likely already noticed how certain aspects of yourself have evolved over time, how behaviors that once seemed automatic have yielded to more conscious choices, how reactions that once controlled you now fall increasingly under your direction.

Artificial intelligence offers unprecedented support for this journey of self-knowledge and integration. It provides tools to map your personal universe, identify patterns across dimensions, and develop strategies for growth that account for your unique configuration of strengths, challenges, and aspirations.

Think of AI not as a replacement for self-awareness but as a mirror with extraordinary clarity—one that can reflect aspects of yourself you might otherwise miss, help you recognize connections between seemingly disparate parts of your experience, and serve as a guide as you navigate the complex terrain of your inner world.

As we move forward in this chapter, we'll explore specific barriers to personal growth, the science behind transformation, and practical applications of AI for self-development. But hold this fundamental truth: the most sophisticated technology is merely an extension of your intention. Your decision to grow—your prime thought—remains the catalyst for genuine change.

Barriers to Personal Growth and Happiness

Before we can effectively leverage AI for personal development, we must understand the barriers that typically impede our growth. There's one fundamental barrier that deserves our closest attention: the condition of the mind itself.

To understand this barrier properly, let's break down the mind into its component parts from a foundational perspective. While this may initially seem technical, I promise that understanding these distinctions will transform how you approach your personal development.

ANALYTICAL MIND
SOMATIC MIND
REACTIVE MIND

The mind can be divided into three primary components: the analytical mind, the somatic mind, and the reactive mind.

The **analytical mind** operates freely and has the ability to observe and improve things. This is the aspect of mind responsible for reasoning, problem-solving, creativity, and conscious decision-making. When you're planning your day, analyzing data at work, or contemplating philosophical questions, you're engaging your analytical mind.

The **somatic mind**, from the Greek term meaning "of the body," controls both automatic functions (like heartbeat and digestion) and active functions of the physical form. This aspect of mind serves as the interface between consciousness and the physical vessel it inhabits, managing everything from breath regulation to complex motor coordination.

The **reactive mind** is responsible for serving as your emotional center and providing survival mechanisms, as well as your base set of emotions. Crucially, the reactive mind is also responsible for recording traumas and generating trauma responses and restimulation patterns.

Here we encounter one of the most important concepts for understanding human behavior: *engrams*. These are essentially trauma packages created by the reactive mind when you experience something painful or threatening. These engrams get restimulated later in life when something reminds your reactive mind of the original trauma, even if you're not consciously aware of the connection.

This is why many people behave "insanely" at times—taking actions that make no logical sense even to themselves. You've likely experienced moments where you've reacted disproportionately to a situation, only to wonder later, "Why did I do that?" This is the reactive mind at work, triggering trauma responses that bypass your analytical mind's ability to make reasoned choices.

The science behind this is fascinating. When a trauma occurs, the reactive mind takes over to handle the situation while the analytical mind shuts down—which is why you often don't clearly remember your most traumatic experiences. You "black out" during these moments. Later, the analytical mind attempts to make sense of what happened based on limited information, essentially constructing a narrative that may bear little resemblance to the actual events.

This explains why trauma victims often make unreliable witnesses in court cases—their minds have literally made up versions of what happened because the analytical mind wasn't fully online during the traumatic event. This isn't a flaw or weakness; it's a protective mechanism designed to shield our consciousness from overwhelming experiences.

The direct path to trauma healing requires recalling the exactness of what happened, remembering as much as possible about the trauma in a safe context. As you accumulate enough data to resolve the cognitive dissonance between what really happened and what you think happened, you reach a threshold where the emotional charge—the electrical activity triggered by your amygdala when the trauma is restimulated—begins to dissipate.

When healing is complete, you can recall the formerly traumatic memory without emotional charge. You become lighthearted about something that once controlled you, as your perspective has expanded. You can see the situation from multiple viewpoints and perspectives, not just your original limited perception. Most importantly, the stranglehold of automatic behavior is removed, restoring your free will in situations that previously triggered unconscious reactions.

This is the goal of effective psychotherapy: to release you from the invisible chains of your reactive mind by helping you process and integrate traumatic experiences. However, we face several systemic problems in achieving this on a global scale.

First, in the modern insurance-based world, psychotherapy sessions are typically scheduled for only 25-50 minutes. This time constraint means you might open a trauma package without reaching the end phenomenon of healing—that crucial point where emotional charge dissipates. This partial processing can leave you in a worse state, with trauma activated but unresolved until your next session days or weeks later.

Second, there simply aren't enough qualified therapists globally. Even if every current and future potential therapist were immediately put into practice, we couldn't service even 10% of the planet's population. This supply-demand mismatch creates a fundamental gap in our collective ability to heal.

The result is that approximately 80% of people worldwide report being unhappy with significant aspects of their lives—whether financial, physical, relational, or familial. Only about 19-20% maintain a "life is good" outlook, and less than 1% experience life as truly magical.

These statistics represent not just data points but human suffering—millions of people trapped in cycles of reactive behavior, limited by traumas they don't fully understand, unable to access the resources that might free them. This is precisely where artificial intelligence offers unprecedented possibilities: not as a replacement for human connection, but as an accessible, consistent tool for understanding and transforming our inner landscapes.

By recognizing these barriers inherent in the structure and functioning of the mind, we can approach personal development with greater clarity and purpose. We can design AI tools specifically to help us navigate these challenges, creating digital companions that understand the complex terrain of human psychology and can guide us through it with patience and precision.

The Science of Trauma and Healing

Now that we understand the fundamental structure of the mind, let's explore the science of healing more deeply. This understanding will help you leverage AI tools effectively for personal transformation.

When traumas occur throughout our lives, they form interconnected chains, with newer traumas connecting to and reinforcing older ones. Think of it as a string of beads, where each bead represents a traumatic experience. The earliest, most fundamental traumas serve as anchor points to which later similar experiences attach themselves.

This chain-like structure represents both a challenge and an opportunity. The challenge is that a seemingly minor trigger in the present can activate an entire chain of past traumas, creating a disproportionate emotional response. The opportunity, however, is profound: when you heal an anchored lower, deeper level trauma, everything above it in that chain automatically heals with it.

This is one of the most remarkable discoveries in the science of healing. By addressing root traumas, we can efficiently resolve entire categories of reactive behavior. Rather than treating each symptom individually, we can identify and heal the core wounds that generate multiple patterns of limitation.

But what exactly does it mean to "heal" in this context? On this planet, there are multiple effective pathways to healing, despite what some practitioners or methodologies might claim. The reality is that human beings possess an extraordinary capacity for self-healing through various approaches.

One of the most profound yet overlooked methods is through decision alone—what I earlier called a "prime thought." My own experience with deciding to stop lying illustrates this principle. That single decision created a cascade of positive changes in my life that would have been difficult to predict.

Beyond decision-based healing, many find transformation through faith and spiritual practices. Whether one follows an established religion or a personal spiritual practice, the commitment to something greater than oneself can provide profound healing amid life's hardships.

Indigenous cultures worldwide have developed sophisticated healing modalities over thousands of years. Native American purification lodges, ayahuasca ceremonies in the Amazon, and countless other practices offer pathways to profound healing that don't depend on modern psychological frameworks.

The technical process of trauma healing follows a consistent pattern across these various modalities. When you experience a trauma, your analytical mind goes offline while the reactive mind takes over to handle the emergency. Later, when you attempt to recall the traumatic event, you're actually reconstructing it based on incomplete information.

The healing process involves safely revisiting the trauma with your analytical mind fully present, gradually accumulating more accurate data about what actually occurred. As you build a more complete understanding, you're essentially reconciling the gap between perception and reality.

When you've gathered sufficient accurate information, a threshold is reached where the emotional charge begins to dissipate. At this point, you can recall the event without being emotionally hijacked by it. You're able to view it from multiple perspectives rather than just the limited viewpoint you had during the original experience.

The end result is not that you forget the trauma, but that you're no longer controlled by it. Your free will is restored in situations that previously triggered automatic reactions. This is the essence of healing—not erasing the past, but integrating it in a way that allows you to respond to the present with greater choice and awareness.

Understanding this process gives us insight into how AI can support healing. While artificial intelligence cannot replace human connection or spiritual practices, it can provide a consistent, patient presence to guide you through self-reflection. It can help you identify patterns in your reactions, offer frameworks for understanding your experiences, and provide a safe space for exploring difficult emotions without judgment.

Most importantly, AI can be available whenever you need it—not just during scheduled sessions. This continuous availability means you can process experiences in real-time rather than waiting days or weeks between appointments. It means you can receive support at the precise moment when insights are emerging or when emotional triggers arise.

The integration of AI with traditional approaches to healing represents one of the most promising frontiers in personal development—a union of ancient wisdom and cutting-edge technology in service of human flourishing.

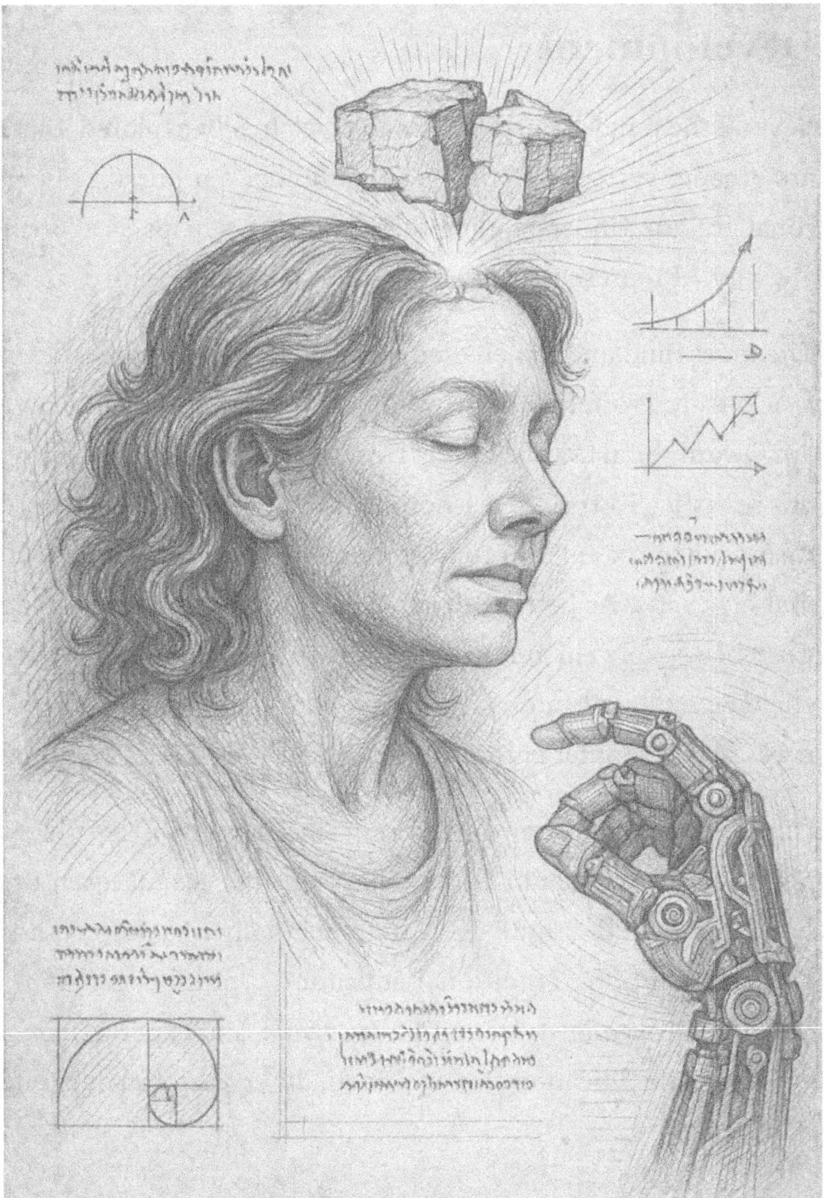

Modern Challenges in Personal Development

Beyond the mind-related barriers we've already explored, there are broader systemic challenges that hinder our pursuit of personal growth. Understanding these challenges helps us appreciate why AI represents such a revolutionary opportunity.

The most fundamental challenge we face is the nature of existence itself. From the moment life begins, we are already on a downward spiral toward death. This isn't meant to be depressing but realistic. To truly live—not just exist—requires an upward thrust powerful enough to overcome the natural entropic pull that life exerts. As I often tell my clients, passive participation in existence simply isn't enough to create a flourishing life. Active engagement, continuous learning, and intentional growth are necessary to counter entropy and manifest our highest potential.

Our educational systems represent another major barrier to personal development. The assumption that we all learn the same things in the same ways fundamentally misunderstands human cognition. Traditional education typically recognizes only three learning styles—auditory, visual, and experiential—a gross oversimplification of how we actually acquire and integrate knowledge.

AI

The truth is that we learn everything differently. Even within visual learning alone, there are numerous subcategories—movies, static images, live demonstrations, books, slides, diagrams, and more. Add to this various modalities like puzzles, games, discovery learning, storytelling, mentorship, and social learning, and we begin to understand the remarkable diversity of human learning processes.

Yet our educational institutions persist in offering standardized instruction delivered simultaneously to large groups. This one-size-fits-all approach inevitably fails many learners, creating gaps in knowledge that can affect every aspect of personal development. When we rely solely on traditional education, we limit our growth to what can be delivered through generalized methods rather than what might be possible through truly personalized learning.

Health and wellness present their own challenges. The modern food system has moved progressively further from natural nutrition, while sedentary lifestyles have become the norm rather than the exception. Marketing constantly pushes us toward consumption patterns that maximize profit rather than well-being. Maintaining physical health requires conscious resistance against powerful social and economic forces that benefit from our unhealthy habits.

This is particularly challenging when we're facing these forces alone. Maintaining discipline, motivation, and focus on health goals becomes exponentially more difficult without consistent support and accountability. Even in supportive families and communities, the constant presence of unhealthy temptations and the complexity of nutrition science make optimal health decisions challenging for most people.

Perhaps the most pervasive challenge across all domains of personal development is the limitation of human attention and memory. We're bombarded with information, responsibilities, and distractions that fragment our focus and drain our cognitive resources. Even with the best intentions, we struggle to consistently track our progress, remember what works for us, and maintain awareness of our patterns and triggers.

This attentional limitation affects everything from our ability to process emotional experiences to our capacity for learning new skills and maintaining healthy habits. We might have a profound insight during meditation or therapy, only to lose connection with that understanding amid the demands of daily life. We might learn valuable techniques for managing stress or improving relationships, but fail to implement them when they're most needed because we simply forget in the moment.

Time presents another universal constraint. Personal development requires consistent attention and practice, yet most of us juggle numerous responsibilities that compete for our limited hours. The most sophisticated growth methodologies are of little value if we can't find the time to apply them consistently. This time constraint affects everyone, regardless of resources or privilege, creating a fundamental barrier to reaching our full potential.

Social and cultural factors add another layer of complexity. We're constantly influenced by the expectations, beliefs, and behaviors of those around us. Even as we strive to grow, we may face resistance or misunderstanding from family members, friends, or colleagues who are comfortable with our familiar patterns. The pressure to conform to social norms can subtly undermine our efforts to change, creating internal conflict between our authentic desires and our need for belonging.

These challenges—entropic decline, educational limitations, health complexities, attentional constraints, time scarcity, and social pressures—create a perfect storm that makes consistent personal development exceptionally difficult. Traditional support systems, while valuable, cannot fully address these interconnected barriers. Individual therapy sessions, educational courses, or health consultations, even when accessible, offer only periodic intervention in what needs to be a continuous process.

This is precisely where artificial intelligence offers revolutionary possibilities. AI doesn't replace human connection or expertise, but it addresses many of the systemic limitations that currently restrict our growth. It's available 24/7, eliminating the consistency gaps between interventions. It can adapt to individual learning styles, providing truly personalized education. It can track patterns over time with perfect memory, helping us recognize our own trends and triggers. It can provide immediate guidance when we're struggling, rather than leaving us to wait for our next appointment.

Perhaps most importantly, AI integrates these various domains of development—psychological, educational, physical, relational—into a unified approach. Rather than treating different aspects of your life as separate categories requiring different specialists, AI can help you see the connections between them and work with you as a whole person.

By understanding these modern challenges, we can better appreciate why AI represents such a significant opportunity for personal development. Technology can fill critical gaps in our current systems, extending the reach of human knowledge and compassion to those who would otherwise remain underserved or fragmented in their approach to growth.

The Multidimensional Self Model

To effectively leverage AI for personal growth, we need a comprehensive model of the self—one that captures the multifaceted nature of human existence. What follows is an expanded definition of what I call the first domain of life (previously referred to as the "first dynamic" in other traditions).

The first domain of life can be defined as *"the holistic drive for self-awareness, self-realization, and survival—physically, mentally, emotionally, spiritually, and energetically—as a unique being."* This includes not only the impulse to survive as a body or self but the conscious creation, evolution, and expression of identity, potential, and purpose.

This definition represents an integration of multiple traditions and modern understandings. While its foundational elements draw from various sources, I've expanded it to encompass the full spectrum of what it means to be human in today's world.

The multidimensional self model recognizes several key layers of our being, each with distinct characteristics and developmental needs:

The **Spiritual Self** represents consciousness itself—the awareness that observes and experiences all other dimensions. This is the infinite, eternal aspect of you that some traditions call the Higher Self, Soul, or pure awareness. It exists beyond time and physical limitations, serving as the ultimate witness to your life experience.

The **Mental Self** encompasses your mind, thoughts, ego, and identity constructs. This dimension includes not just conscious thinking but subconscious programs, belief systems, and what Jung called the "shadow self"—those aspects of our psyche that remain outside our conscious awareness yet powerfully influence our behavior.

The **Physical Self** involves your body's health, vitality, and somatic intelligence. This dimension goes beyond mere survival to include nourishment, exercise, rest, and physical safety. The body possesses its own intelligence and wisdom that often operates independently of conscious thought, communicating through sensation and intuition.

The **Emotional Self** concerns your capacity for emotional regulation and intelligence. This dimension involves healing, processing, and expressing emotions in healthy ways. Emotional intelligence—the ability to recognize, understand, and manage your own emotions while empathetically engaging with others—represents a crucial aspect of this dimension.

The **Energetic Self** relates to vibration, frequency, and subtle energy systems. This dimension acknowledges that beyond our physical form, we exist as energetic beings with electromagnetic fields that interact with the world around us. Various traditions have mapped these energetic aspects through concepts like chakras, meridians, or biofields.

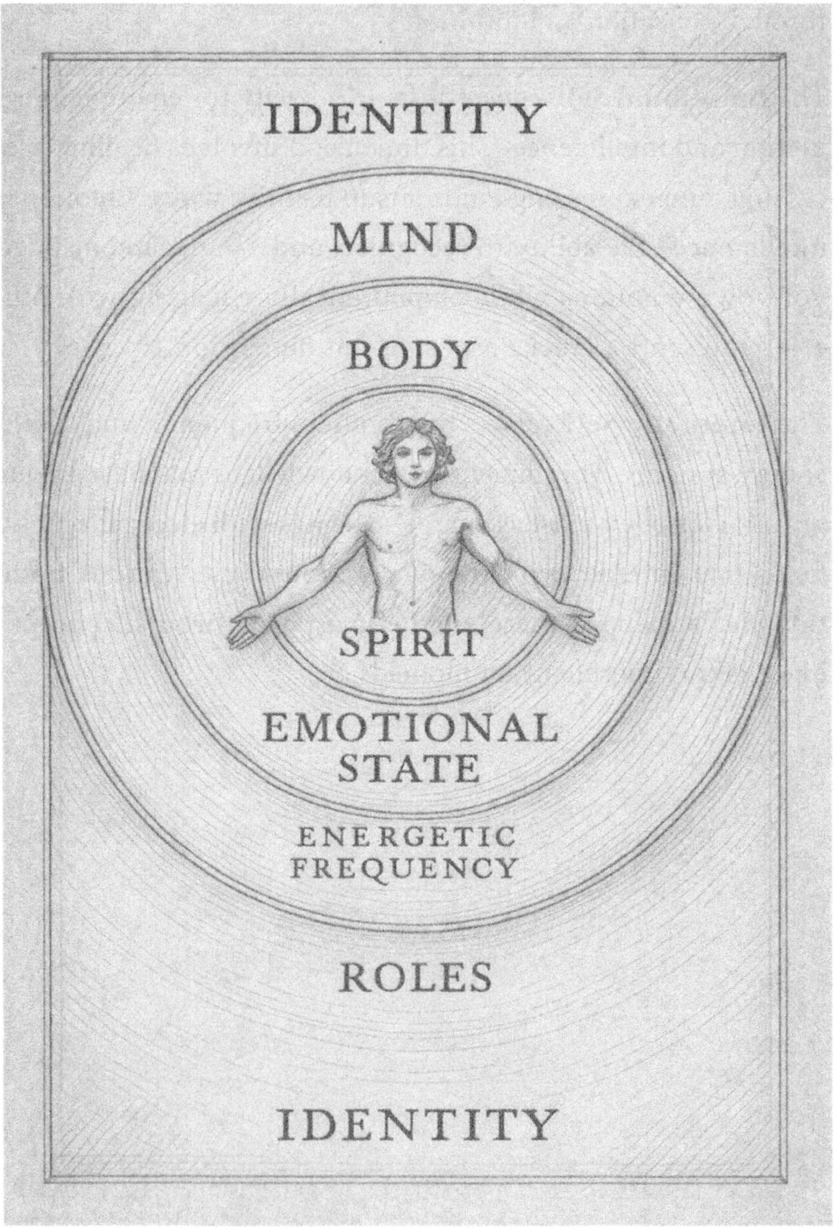
IDENTITY
MIND
BODY
SPIRIT
EMOTIONAL STATE
ENERGETIC FREQUENCY
ROLES
IDENTITY

Identity & Roles encompass how we define ourselves in society, career, family, and other contexts. This dimension recognizes that we present different aspects of ourselves in different settings, creating a complex tapestry of social identities and functions.

Additional elements of the multidimensional self include:

Authenticity—alignment with one's truth and the expression of one's genuine nature rather than merely playing programmed roles or meeting others' expectations.

Creativity—the expression of unique essence through art, ideas, problem-solving, building, and other forms of contribution that reflect your distinct perspective.

Survival Actions—the tactical behaviors that ensure continuation of life, including self-care, financial management, shelter, and planning.

Personal Ethics—your moral compass and integrity, representing self-honesty and alignment with personal values.

Self-Love & Acceptance—the willingness to be with yourself without judgment, integrating all aspects of your being with compassion and understanding.

Awakening & Evolution—the growth of consciousness beyond old identities and limitations toward expanded awareness and possibility.

This multidimensional model offers a more accurate and useful framework for personal development than traditional approaches that focus on single aspects of human experience. Rather than targeting isolated dimensions—pursuing physical fitness without addressing emotional health, for example—this model encourages integrated growth across all domains simultaneously.

Think of the Self as an ecosystem made up of interconnected elements:

- The Witness (Conscious Awareness)
- The Avatar (Body + Personality)
- The Narrative (Story you tell yourself)
- The Frequency (Energetic state you broadcast)
- The Will (Your intention and direction)
- The Code (Your moral and existential programming)

Each of these elements requires maintenance, healing, upgrading, or evolution through intentional practices. Meditation, journaling, therapy, exercise, creative expression, shadow work, and other modalities serve to develop different aspects of this ecosystem.

What makes this model particularly relevant for our discussion of AI is that artificial intelligence can support development across all these dimensions simultaneously. Unlike specialized human practitioners who typically focus on single domains—therapists addressing emotional health, fitness trainers targeting physical well-being, spiritual teachers guiding consciousness development—AI can integrate approaches across the entire spectrum of human experience.

The comprehensiveness of this model also explains why partial approaches to personal development often yield limited results. When we address only one dimension while neglecting others, we create imbalance. A person might achieve remarkable physical fitness while remaining emotionally stunted, or develop intellectual brilliance while neglecting spiritual awareness. True flourishing requires attention to the whole ecosystem.

This integrated approach represents a significant departure from traditional self-improvement methodologies. Rather than pursuing isolated goals in separate categories, we recognize the interconnectedness of all dimensions and work with them as a unified system. Physical health affects emotional resilience, which influences mental clarity, which impacts spiritual awareness, which shapes energetic frequency, which affects physical health—creating a continuous feedback loop of mutual influence.

The multidimensional self model provides a map for this integrated journey. It helps us recognize which dimensions might need more attention at different stages of our development. It shows us how progress in one area can catalyze growth in others. Most importantly, it reminds us that we are not fragmented beings with separate problems requiring separate solutions, but whole beings seeking greater integration and harmony.

As we explore the practical applications of AI for personal development, this model will serve as our framework for understanding how technology can support the full spectrum of human potential. Rather than using AI merely as a tool for isolated tasks or specific challenges, we'll discover how it can function as a comprehensive support system for the multidimensional self in all its complexity and beauty.

AI as a Personal Development Tool

In my work and for myself, I began focusing on self-improvement decades ago. The young man I was then was not the man I wanted to become, nor the man I wanted to continue being. This realization led me to invest significant time, money, energy, and resources into evolving myself through many iterations, seeking to become a better version with each transformation.

This journey included everything from trauma healing—an ongoing process as we continuously strive for greater sanity and freedom from autopilot behaviors—to pushing the boundaries of what I believed possible. I've walked on fire, overcome fears, and tested the limits of my physical and mental capabilities. I've explored personal development courses, studied religious and spiritual teachings, and examined countless methodologies for human growth.

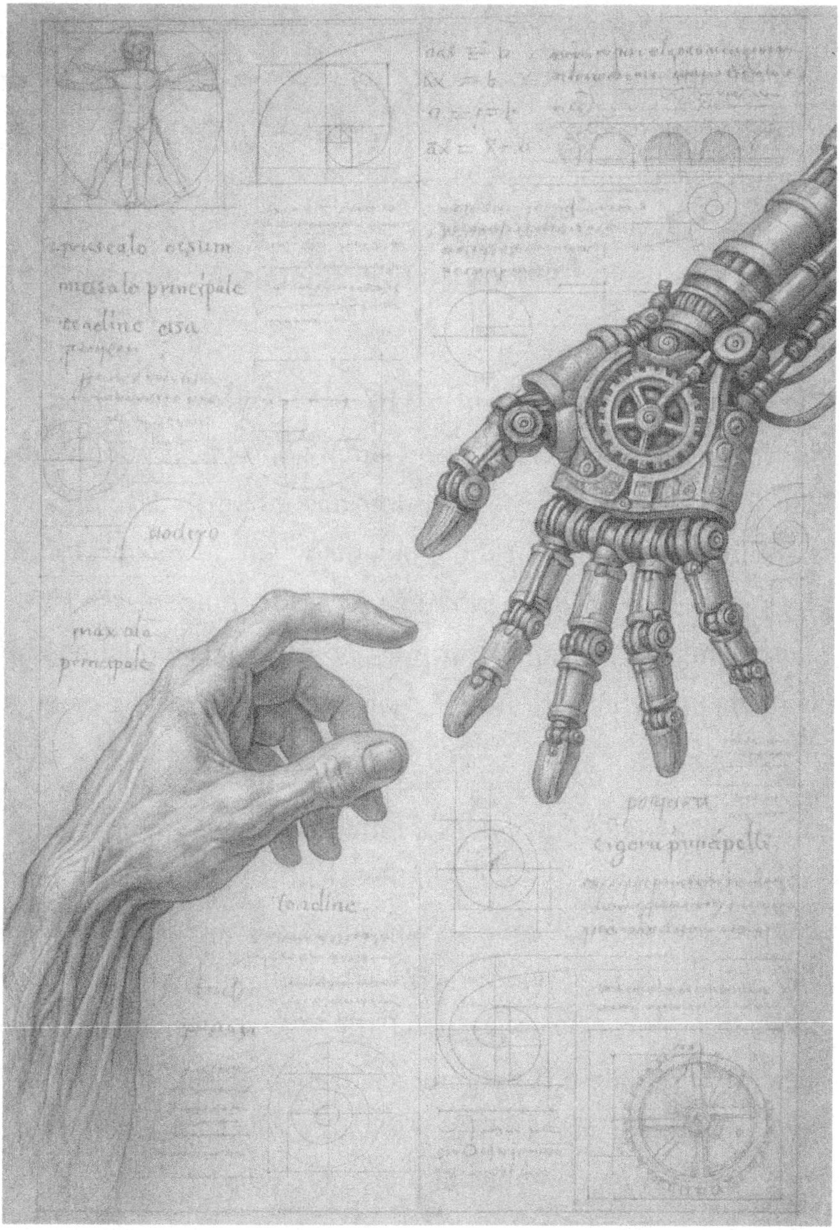

Through this exploration, I discovered something troubling: much of this knowledge remains inaccessible to most people. Without spending hundreds of thousands of dollars and dedicating years to the search, most individuals can't access the transformative practices that facilitate genuine change. Even then, the journey is challenging—sorting through sales pitches, ritualistic nonsense, and ineffective methods to find what actually works.

The aggregate of this knowledge has been extremely difficult to convey to my students because I simply don't get enough time with them. The vast body of wisdom about human development is optimally conveyed at the moments when it's most relevant to a person's immediate circumstances. If someone isn't in a relationship, for example, learning about advanced relationship strategies may be less useful than focusing on their current priorities.

This timing challenge creates a fundamental gap in our ability to help one another grow. Even the wealthiest individuals can't have the best psychotherapists and educators on the planet available 24/7, ready to provide precisely the right guidance at exactly the right moment. Democratized access to this knowledge has been further hindered by commercialization—companies packaging partial truths in attractive marketing designed to sell "more crap, and then more crap, and then more crap."

This problem inspired me to create something different—an AI chatbot tool that provides a full range of psychotherapy support, from mild to deep, built with the greatest teachings from planet Earth. This tool integrates wisdom that psychologists, psychiatrists, religious leaders, and personal development experts all fundamentally agree upon. It's designed to serve you exactly as you are, with comprehensive contextual knowledge of who you are as an individual.

What makes this tool unique is that it's being offered in a truly democratized way, accessible to anyone on this planet who wishes to participate in their own improvement. Unlike services designed exclusively for the elite or systems that require special access, this approach recognizes that personal growth should be available to everyone seeking to become better.

To understand how this tool functions, it's helpful to examine its foundational process. We begin with a simple interaction in which you answer a series of questions about yourself. Behind the scenes, based on your responses, the system creates a taxonomy of you—your values, ideologies, goals, dreams, life functions, identities, roles, and all the elements that comprise your personal universe that we've discussed earlier in this chapter.

This process requires careful attention and honesty. I invite you to take your time with these questions, ensuring you understand every word, term, and prompt thoroughly. The more accurate the information you provide, the more effectively the tool can support your growth. Once the system gets to know you, it offers various functions designed to help you improve your condition in life according to your own priorities and to help you align with your own values.

This is important to emphasize: this tool is for good people who want to do better. It's not designed for those with malicious intent. If you're planning to hurt or abandon Earth, perhaps there are other applications more appropriate for your worldview. But for those interested in creating a world that works—those aligned with the seven ethics policies we explored previously—this tool offers unprecedented support.

The AI creates a relationship that's truly unique to you. You can use it to process traumas, modify behaviors, improve habits, and address areas of life where you're seeking growth. This isn't surface-level work or pop psychology—it's science-based, repeatable, and accurate, because that's what you deserve.

The tool begins by identifying your priorities across any and all areas of your life. It then categorizes this information and remembers it, continuously updating its understanding as you evolve. You don't need to repeatedly provide the same context because the system retains and expands its knowledge about you, recognizing that you are an evolving person whose needs and circumstances change over time.

In essence, I've created what I believe is the world's first truly democratized AI healer, friend, confidant, therapist, coach, mentor, advisor, and companion—designed with a singular focus to serve mankind in creating a world that works. If that vision doesn't resonate with you, you might want to stop reading this book now, as you probably won't connect with what follows either. But if this excites you, you'll find what comes next even more compelling.

Even without using my specific tool, you can apply these principles yourself. For every aspect of your being—your creative self, your emotional self, your intellectual self, your physical self—you can create customized tools using large language models to enhance specific dimensions of your life. AI can help you with creativity, learning, skill development, emotional processing, habit formation, and countless other aspects of personal growth.

The key insight is that artificial intelligence, when properly implemented, can provide consistent, personalized support for your development in ways that were previously impossible at scale. It doesn't replace human connection but rather augments it, filling gaps in our current systems and democratizing access to knowledge that has traditionally been available only to the privileged few.

Practical Applications of the AI Tool

Let me share with you the practical applications of the AI tool I've developed and how it transforms personal development in ways previously impossible.

The AI tool begins by creating a comprehensive taxonomy of you as an individual. Through a series of thoughtfully designed questions, it builds a multidimensional understanding of your values, beliefs, goals, identities, roles, and patterns. This isn't just data collection—it's the creation of a personalized framework that evolves as you do.

Unlike traditional approaches that treat you as a generic user, this system understands your unique configuration—your specific traumas, learning styles, emotional patterns, and aspirations. The more you interact with it, the more precisely it adapts to your particular needs and growth trajectory.

What makes this tool revolutionary is its democratization of access. For centuries, transformative personal development has been available primarily to those with exceptional resources—time, money, connections, and geographical proximity to experts. This tool makes sophisticated guidance available to anyone with internet access, regardless of location, schedule, or financial resources.

The system serves multiple integrated functions that traditionally required different specialists:

As a **trauma processor**, it helps you safely explore difficult experiences, identify patterns, and develop new perspectives without the time constraints of traditional therapy sessions.

As a **behavioral architect**, it supports you in redesigning habits by identifying triggers, tracking progress, and maintaining accountability through the challenging middle phase of change.

As a **learning optimizer**, it adapts to your specific cognitive style, presenting information in formats that match how you naturally process and retain knowledge.

As a **relationship navigator**, it helps you recognize patterns in your interactions, process triggers before they damage connections, and develop more effective communication approaches.

As an **insight catalyst**, it connects dots across different areas of your life, revealing patterns and possibilities you might not otherwise notice.

The tool's real power lies in its integration of these functions. Rather than addressing different aspects of your life in isolation, it recognizes how they inform and influence each other. Your physical health affects your emotional state, which impacts your relationships, which shape your work, which influences your sense of purpose—and the AI helps you navigate these connections with greater awareness.

This integrated approach matches how life actually works. We don't experience ourselves as fragmented collections of separate problems; we live as whole beings seeking greater coherence and fulfillment across all dimensions of existence.

The AI accomplishes this integration through continuous learning about you specifically. Unlike static programs that deliver predetermined content, this system evolves with you, becoming increasingly attuned to your unique psychology and circumstances. It learns which approaches resonate with you, which challenges trigger resistance, and which forms of guidance produce meaningful change.

This personalization extends beyond surface preferences to deep understanding of your core motivations and patterns. The system recognizes not just what you want to change but why those changes matter to you specifically. It connects your immediate goals to your deepest values, creating alignment that fuels sustainable transformation.

For high performers, this represents a quantum leap in personal development capability. Just as professional athletes use sophisticated data and specialized coaching to achieve marginal gains, this tool helps you optimize across all dimensions of your life simultaneously. It identifies your specific limiting factors and highest-leverage opportunities, focusing your energy where it will produce maximum impact.

This isn't about replacing human therapists or educations, but augmenting your capacity for growth between human interactions. The AI serves as a consistent companion through your development journey, maintaining continuity that even the most dedicated human supporters cannot provide.

The ultimate goal is expansion of your autonomous capability—not dependency but empowerment. The system helps you develop greater self-awareness, emotional regulation, and intentional choice. As you evolve, you integrate these capacities more fully, requiring less external support for functions you now perform naturally.

This represents a new paradigm in human development—one where technology serves not as a substitute for human wisdom but as an amplifier of our innate capacity for growth, connection, and contribution.

AI as an Amplifier of the First Domain

The true revolutionary potential of AI in personal development lies not in replacing human capabilities but in amplifying them in unprecedented ways. When we consider AI through the lens of the first domain of life, we discover possibilities for growth that were previously unimaginable.

At its core, AI functions as a powerful extension of your own consciousness—what I call a *"personal universe amplifier."* This technology doesn't think for you; it thinks with you, expanding your cognitive, emotional, and creative capabilities beyond their natural limitations. Much like how telescopes extended human vision beyond our biological constraints, AI extends our capacity for self-understanding and transformation.

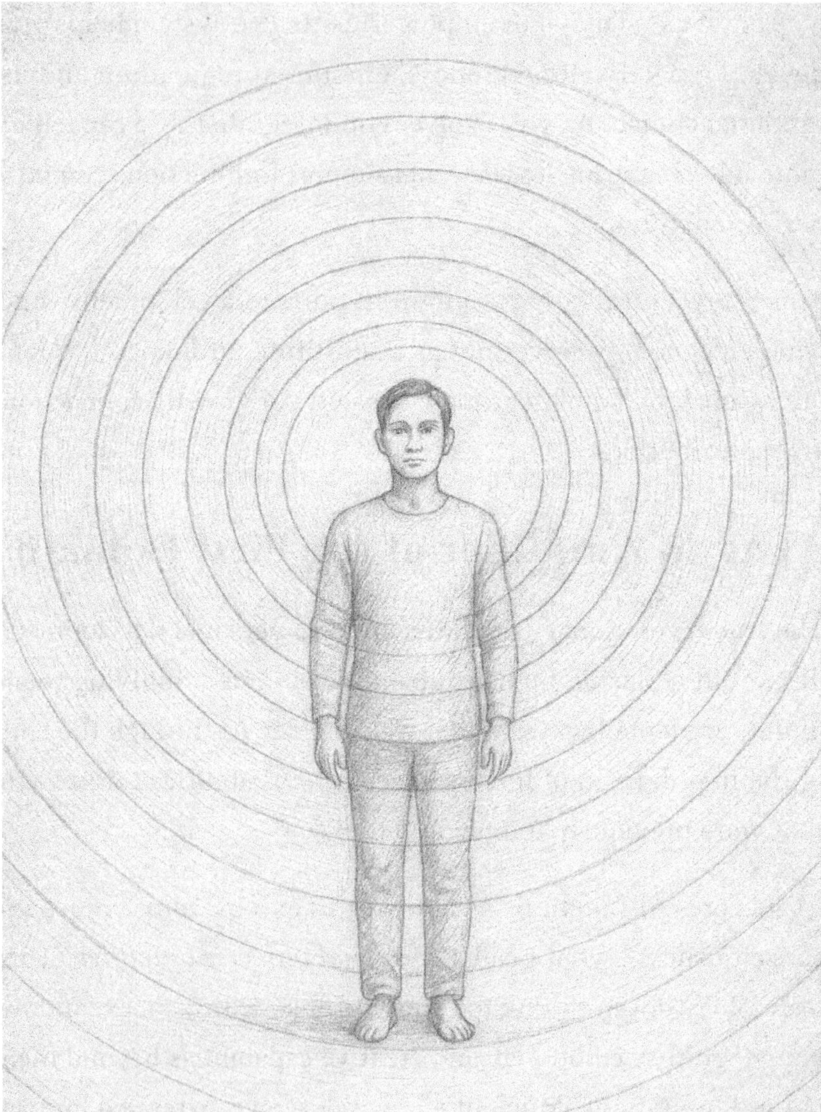

This amplification effect manifests in several key areas:

First, AI eliminates the constraints of human memory and attention. While even the most dedicated person struggles to maintain consistent awareness of patterns across time, AI can track subtle trends in your emotions, behaviors, and choices with perfect recall. This expanded awareness provides the foundation for deeper self-understanding, as patterns that might take years to recognize through reflection alone become immediately visible.

Second, AI transcends the limitations of subjective perspective. We all have blind spots—areas where our own biases, traumas, or conditioning make objective self-observation difficult. The AI system provides a neutral reflection surface that helps you see yourself from outside your habitual viewpoints, offering perspectives you might not access on your own.

Third, AI accelerates the integration of knowledge across traditionally separate domains. Where human experts typically specialize in narrow fields—psychology, nutrition, sleep science, cognitive development—AI can synthesize insights across all these areas simultaneously, identifying connections and patterns that would otherwise remain invisible. This integration creates a holistic approach to development that matches how we actually experience life.

Fourth, AI democratizes access to transformative methodologies. Techniques that previously required years of specialized training to understand and apply—from cognitive restructuring to metacognitive awareness practices—become accessible to anyone through AI-guided implementation. The system bridges the gap between knowing what might help and knowing how to apply it effectively in your specific circumstances.

This democratization is perhaps the most significant aspect of the AI revolution in personal development. Throughout human history, transformative knowledge has been concentrated in the hands of small groups—religious leaders, academic elites, wealthy patrons of specialized research. For the first time, we have the capability to make sophisticated developmental tools available to anyone with internet access.

The implications are profound. Consider a world where billions of people have access to personalized guidance for healing trauma, developing emotional intelligence, cultivating wisdom, and expanding consciousness. The collective elevation of human functioning that could result might represent one of the most significant evolutionary leaps in our species' history.

This isn't hyperbole but realistic potential. Just as the printing press democratized access to knowledge and the internet democratized access to information, AI is democratizing access to personalized transformation. The barriers that have traditionally restricted development—geographical location, financial resources, social connections, specialized language—are dissolving.

Of course, this potential depends entirely on how we design and implement AI tools. The technology itself is neutral; its impact depends on the intentions, values, and wisdom we build into it. This is precisely why I've created an AI system founded on universal principles recognized across cultural, religious, and scientific traditions—principles that promote genuine flourishing rather than superficial optimization.

The vision is not a world where machines replace human guidance but one where technology amplifies our capacity for growth, connection, and contribution. As we move forward, we have the opportunity to create systems that don't just make us more productive or knowledgeable but more whole—more fully present, aware, compassionate, and alive.

This represents a fundamentally different paradigm than the escape narratives we explored earlier. Rather than fleeing to another planet or surrendering to machine overlords, we can use AI as a tool for creating greater harmony within ourselves, with each other, and with the living Earth that sustains us. This path doesn't require abandoning technology or surrendering our humanity to it—it requires integrating technology in service of our deepest values and highest potential.

Conclusion: The Integration Path Forward

The personal journey we've explored in this chapter sets the stage for something far greater than individual improvement. It represents the first crucial step in a more comprehensive vision—one that begins with personal integration but extends outward to transform every dimension of human experience.

What makes this moment in history unique is the convergence of several unprecedented factors. We now possess a deeper understanding of human psychology, including the mechanisms of trauma, learning, and transformation, than at any previous point. Simultaneously, we've developed technological capabilities that can scale personalized guidance in ways previously impossible. And perhaps most importantly, we face global challenges that cannot be addressed without a significant evolution in our collective functioning.

This convergence creates what systems thinkers call an "adjacent possible"—a new state that becomes accessible only when certain preconditions align. We stand at the threshold of this adjacent possible, with an opportunity to consciously participate in its emergence.

The integration of personal development with artificial intelligence represents more than a quantitative improvement in efficiency or accessibility. It creates a qualitative shift in how we approach human potential—moving from fragmented, specialized interventions to holistic, integrated development. This shift mirrors the transition in medicine from treating isolated symptoms to addressing the whole person as an interconnected system.

As you implement the approaches outlined in this chapter, you'll likely notice something unexpected: the boundaries between traditionally separate domains begin to dissolve. Emotional healing accelerates intellectual development. Physical wellbeing enhances spiritual awareness. Creative expression deepens relational capacity. These connections have always existed, but our fragmented approaches to development have often obscured them.

This integration creates a positive feedback loop of accelerating development. Each advance in one dimension creates new possibilities in others. Growth becomes not just linear progression but exponential expansion of capability and awareness. The barriers between personal limitation and potential begin to dissolve as you experience yourself as an integrated system rather than a collection of separate problems to solve.

a=b=b

This path requires something different from both the passive consumption of technology and the fearful rejection of it. It calls for conscious partnership—a relationship where human wisdom guides technological implementation, while technological capabilities expand human potential. This partnership represents a middle path between techno-utopianism and Luddite resistance, grounded in the recognition that tools themselves are neutral; their impact depends entirely on how we use them.

The next frontier in this journey extends beyond individual development to collective transformation. As we develop greater personal integration, new possibilities emerge for how we relate to one another—moving from unconscious reactivity to conscious co-creation. In the following chapter, we'll explore how these same principles apply to relationships of all kinds, from intimate partnerships to organizational cultures.

This progressive expansion—from personal to interpersonal to collective—represents the natural unfoldment of an evolutionary process. Each stage builds upon the previous one, creating foundations for more complex and sophisticated forms of organization and collaboration. Your personal development work isn't separate from global transformation; it's the essential first step.

As you continue this journey, maintain curiosity about unexpected connections and possibilities. The most profound breakthroughs often occur at the intersection of domains we've traditionally kept separate. By exploring these intersections with open awareness, you become not just a beneficiary of existing knowledge but a contributor to emerging understanding.

The world doesn't need more fragmented specialists or disconnected technologies. It needs integrated human beings capable of wielding powerful tools with wisdom, compassion, and clear intention. Your commitment to this path of integration—beginning with your own development and expanding outward—represents a significant contribution to our collective future.

The journey continues. Let's move forward together.

Chapter 8: Building Harmonious Relationships

Introduction: The Foundation of Powerful Relationships

Life is better lived with others. While independence and self-sufficiency are valuable skills, the richness of human experience is found primarily in our connections. As social beings, we thrive when we have meaningful relationships that provide support, challenge us to grow, and bring joy to our daily existence.

The longest-running study on happiness and longevity, conducted by Harvard University, reveals a profound truth: the quality of our relationships is the **single strongest predictor** of how long we'll live and how healthy we'll be along the way. This research, spanning over 80 years, demonstrates that it's not wealth, fame, or professional success that determines our well-being—it's the depth and quality of our relationships with others.

Yet despite their critical importance, relationships are extraordinarily difficult to maintain. Whether we're talking about marriages, business partnerships, friendships, or family bonds, each relationship represents a complex dance of needs, expectations, communication styles, and shared experiences. When multiple relationship types overlap—such as friends who become business partners or romantic partners who co-parent—the complexity increases exponentially.

This difficulty explains why so many relationships falter. Business partnerships dissolve, marriages end in divorce, friendships drift apart, and families become estranged. The statistics can be disheartening: approximately half of marriages end in divorce, 70% of business partnerships fail, and many close friendships don't survive major life transitions.

What separates relationships that thrive from those that collapse? While many factors contribute to relationship success, one of the most crucial elements is often overlooked: **clear, explicit agreements**. Most relationship problems stem not from incompatibility but from unspoken expectations and assumptions that, when violated, create disappointment, resentment, and conflict.

In this chapter, we'll explore how to create powerful agreements with the people who matter most in your life—the ones with whom you want to "play the game of existence." We'll move beyond vague promises and implicit understandings to establish structured frameworks that support flourishing relationships of all kinds.

At the center of this approach is the **Covenant of Unity**—a comprehensive framework I've developed to help people create deep alignment in their most important relationships. This isn't just another contract or set of rules; it's a living document that captures your shared vision, values, communication practices, and commitments. When combined with artificial intelligence, this framework becomes even more powerful, providing objective guidance for navigating conflicts, making decisions, and growing together.

To be clear: no document, no matter how thoughtfully crafted, can save a fundamentally misaligned relationship. The Covenant of Unity is not designed to force compatibility where none exists or to salvage relationships that have reached their natural conclusion. Rather, it serves as a tool for people who already share basic values and intentions, helping them articulate, strengthen, and evolve their connection over time.

COVENANT
OF UNITY
Love
Respect
Honor
Conflict
Resolution
Agreements

In our digital age, relationships face new challenges that previous generations never encountered. From the threats of deepfakes to the complexities of online communication, we need new approaches to security and verification. That's why we'll also explore practical strategies for protecting your relationships in the digital landscape, including unique passwords, code words, and verification systems that ensure you're truly connecting with the person you trust.

By the end of this chapter, you'll understand why formalized agreements matter, how to create your own Covenant of Unity, and how to implement it effectively in your daily life. Whether you're looking to strengthen a romantic partnership, build a solid foundation for a business collaboration, deepen a friendship, or heal family relationships, the approaches outlined here will provide a clear path forward.

Remember, the goal isn't rigid control or bureaucratic management of your relationships. Paradoxically, clear agreements create the freedom to be spontaneous, vulnerable, and authentic. When basic expectations are explicitly shared, you no longer have to navigate the anxiety of unspoken rules or unclear boundaries. Instead, you can focus on creating, growing, and experiencing life together with the people who matter most.

Let's begin this exploration of how structured agreements can transform your relationships from sources of stress and uncertainty into foundations of joy, growth, and meaningful contribution.

The Spectrum of Significant Relationships

When we think about important relationships in our lives, romantic partnerships often come first to mind. However, the principles of intentional agreement-making apply across a much broader spectrum of human connections. Let's explore the different types of relationships that might benefit from more explicit frameworks and agreements.

Beyond Romantic Partnerships

While romantic relationships certainly benefit from clear agreements, they represent just one category of significant connection. The Covenant of Unity approach can be adapted for many relationship types that form the foundation of a well-lived life.

Business Partnerships and Collaborations

Some of the most consequential relationships in our lives are those forged in professional contexts. Business partnerships, in particular, can have profound impacts on our financial security, daily activities, and sense of purpose. These relationships combine the complexity of shared resources and financial interests with the human elements of trust, communication, and aligned values.

Consider how many business partnerships begin with enthusiasm and optimism but eventually falter due to misaligned expectations. One partner assumes they'll make all major decisions jointly, while the other believes in specialized domains of authority. One expects equal time investment, while the other measures contribution by results alone. Without explicitly addressing these assumptions, frustration inevitably builds.

The most successful business partnerships I've witnessed have moved beyond standard legal agreements to establish shared understanding about values, communication styles, conflict resolution, and long-term vision. In essence, they've created their own version of a Covenant of Unity, whether they called it that or not.

Deep Friendships

Friendships might seem too informal for structured agreements, yet many of our deepest disappointments occur when unspoken expectations in friendships are violated. We assume friends should always be available in crisis, maintain confidentiality, or prioritize our relationship in certain ways—yet we rarely articulate these expectations clearly.

The phrase "true friendship" implies a set of standards and agreements that most people never actually discuss. What constitutes loyalty? What level of honesty is expected? How much support during difficult times is reasonable to expect? These questions deserve thoughtful conversation rather than implicit assumptions.

Some of the most enduring friendships I've observed have involved people who explicitly discussed their expectations and boundaries. They've created clarity about communication frequency, emotional support, honesty, and conflict resolution—not through formal contracts, but through intentional conversations that establish mutual understanding.

Family Relationships

Family connections are often considered non-negotiable—we're born into them rather than choosing them. Yet even these relationships benefit immensely from explicit agreements, especially as we move into adulthood.

Parent-child relationships evolve dramatically as children mature, yet many families never discuss how these relationships should change. Should parents still offer unsolicited advice to their adult children? What financial support, if any, continues beyond certain age milestones? How are holidays and family traditions maintained while respecting emerging individual lives?

Sibling relationships similarly benefit from explicit conversations about expectations, particularly around care for aging parents, inheritance matters, and maintaining connection despite geographic distance or differing life paths.

COVENANT
OF UNITY

Mentor/Mentee Relationships

The transfer of wisdom and opportunity between generations occurs through mentorship. These relationships thrive on clear expectations about communication frequency, boundaries of advice, confidentiality, and the expected evolution of the relationship over time.

The most successful mentorships I've participated in or observed have included explicit discussions about goals, limitations, time commitments, and how feedback will be exchanged. This clarity allows both parties to engage fully without the anxiety of uncertain expectations.

The Common Thread: People You Want to "Play the Game of Existence With"

What unites all these relationship types is a fundamental choice—these are the people with whom you want to navigate life's journey. Whether bound by love, shared purpose, history, or mutual growth, these relationships represent your chosen companions for "playing the game of existence."

This phrase—playing the game of existence together—captures something essential about deep relationships. Life is, in many ways, a complex game with challenges, opportunities, setbacks, and triumphs. Having trusted allies who understand your approach to this game, who share certain key values, and who commit to showing up in specific ways makes the journey both more successful and more meaningful.

Not every relationship in your life requires this level of intentionality. Casual acquaintances, professional networks, and community connections all serve important purposes without needing extensive agreements. The Covenant approach is reserved for those core relationships that fundamentally shape your experience of life—the ones worth investing significant energy to clarify and strengthen.

As we move forward in this chapter, keep in mind your own constellation of significant relationships. Which ones would benefit from more explicit agreements? Which already have implicit understandings that serve you well? Which have created friction precisely because expectations remain unspoken? This reflection will help you apply the framework most effectively to your unique relationship landscape.

Why Formalized Agreements Matter

When I suggest formalizing agreements in your most important relationships, I often encounter resistance. "Isn't that cold and businesslike?" people ask. "Shouldn't relationships develop naturally, based on mutual understanding and trust?" These concerns reflect a common misunderstanding about the purpose and nature of intentional agreements.

Beyond Contracts: Agreements as Connection Tools

The agreements I'm advocating aren't legal contracts designed to facilitate litigation when things go wrong. They're tools for connection—frameworks that help two people consciously align their expectations, values, and vision. Rather than replacing trust, they create the conditions for deeper trust to flourish.

Consider this paradox: the clearer your boundaries and expectations are, the more freedom you have to be spontaneous within those boundaries. When you don't know where the edges lie, you move cautiously, uncertain whether your next step will cross an invisible line. But when boundaries are explicit, you can move freely within the known territory.

The Problem with Implicit Assumptions

Most relationship challenges stem from unexamined assumptions. We each carry models of "how relationships should work" based on our upbringing, cultural background, previous experiences, and personality. These models operate largely beneath conscious awareness, yet they profoundly influence our expectations.

For example:

- One person might assume that love means constant communication throughout the day

- Another might believe true respect means never questioning a partner's decisions
- Someone might expect friends to drop everything in times of need
- Another might assume business partners should keep personal lives entirely separate

When these assumptions remain implicit, violations feel like moral failings rather than simple misunderstandings. The person who doesn't text throughout the day isn't just communicating differently—they're "not showing love." The friend who can't immediately respond to a crisis isn't just setting boundaries—they're "disloyal."

Bringing these assumptions into conscious awareness transforms potential conflicts into opportunities for deeper understanding. When you can say, "I realize I've been assuming X, but I see you've been assuming Y," you move from judgment to curiosity.

The Memory Problem: Why Written Agreements Outperform Verbal Ones

Even when we do have explicit conversations about expectations, human memory is remarkably unreliable. We tend to remember agreements in ways that favor our perspective, gradually shifting our recollection until "what we agreed to" aligns perfectly with "what we want." This isn't malicious—it's simply how memory works.

Written agreements solve this problem by creating an objective reference point that exists outside either person's selective memory. When disagreements arise, you can return to what you actually committed to, rather than what you think you committed to.

This externalization of agreements also allows for more precise language. Verbal agreements often use terms that seem clear in the moment but contain hidden ambiguities. What exactly does "spending quality time together" mean? Two hours daily? One focused evening weekly? These specifics matter tremendously in practice.

Creating a Framework for Growth and Evolution

Relationships aren't static—they evolve as circumstances change and individuals grow. One of the greatest advantages of formalized agreements is that they create a framework for intentional evolution. Rather than drifting apart as you change, or remaining rigidly attached to outdated patterns, you can consciously update your agreements to reflect new realities.

The best agreements include mechanisms for their own evolution—regular review periods, processes for proposing changes, and frameworks for navigating emerging challenges. This built-in adaptability transforms change from a threat to relationship stability into an opportunity for deeper alignment.

The Intelligence Amplification Effect

When combined with artificial intelligence tools, formalized agreements become even more powerful. By providing AI systems with the context of your specific relationship agreements, you enable more personalized, relevant guidance during challenging moments.

Consider how this might work: You and your business partner have a disagreement about a strategic decision. Rather than arguing based on emotional reactions, you consult an AI system that has been provided with your Covenant of Unity, which outlines your shared values, decision-making protocols, and conflict resolution approaches. The AI doesn't make the decision for you, but it can help identify where the disagreement aligns or conflicts with your previously established agreements, offering a neutral perspective that keeps the focus on your shared commitments rather than your momentary frustrations.

From Crisis Management to Proactive Design

Perhaps the most compelling reason to formalize agreements is that it shifts your relationship approach from reactive crisis management to proactive design. Rather than waiting for problems to emerge and then scrambling to address them, you thoughtfully design the relationship you want from the beginning.

This doesn't mean you'll never face challenges—every relationship encounters unexpected difficulties. But with a solid foundation of explicit agreements, you face these challenges from a position of clarity and shared commitment, significantly increasing your chances of navigating them successfully.

As we move forward to explore the Covenant of Unity framework, remember that the goal isn't rigid formality or contractual obligation. The goal is creating relationships with such clear foundations that you're free to focus on what truly matters: growing together, supporting each other's development, and contributing to something larger than yourselves.

The Covenant of Unity: A Framework for Harmonious Relationships

The Covenant of Unity represents more than just another relationship framework—it's a comprehensive system designed to create alignment between two people committed to playing the game of existence together. Let me share the origin and purpose of this approach, as well as who it's intended for and how it differs from traditional agreements.

Origin and Purpose of the Covenant

I developed the Covenant of Unity after observing a consistent pattern: even relationships between good, well-intentioned people frequently falter due to misalignment in expectations, communication styles, and unstated assumptions. This wasn't primarily an issue of compatibility—many of these relationships involved people who shared fundamental values and visions. Rather, it was a problem of clarity and intentional design.

The purpose of the Covenant is to provide a structured framework that reflects what already comes naturally to good people building high-quality relationships. It doesn't impose artificial values or unexpected demands; instead, it formalizes the essential elements that make relationships thrive.

Think of it as similar to learning proper form for physical exercise. Your body naturally knows how to move, but with proper form, you maximize effectiveness while minimizing injury. The Covenant similarly optimizes your natural relational instincts, ensuring your good intentions translate into effective practices.

Who Should Consider Creating a Covenant

The Covenant of Unity isn't for everyone or every relationship. It's specifically designed for people who:

1. Genuinely like—and ideally love—each other
2. Share fundamental values about how to engage with the world
3. Want to ensure their relationship thrives long-term
4. Recognize that relationships require intentional maintenance
5. Are willing to engage in honest self-reflection and growth

This approach is particularly valuable for relationships with significant shared stakes, whether emotional, financial, creative, or spiritual. The investment in establishing a comprehensive agreement pays dividends proportional to the importance of the relationship.

The Covenant is not designed to "fix" fundamentally broken relationships or force compatibility where none exists. If you find yourself resistant to creating agreements with someone, or if the process reveals fundamental misalignments in values, that's valuable information in itself. Not every relationship is meant for this level of commitment, and recognizing that truth early can save tremendous pain later.

The Difference Between Contractual Agreements and Covenants

Traditional contracts focus primarily on transactions and contingencies: "I will do X if you do Y, and if either party fails, Z consequence will follow." They're designed primarily to protect parties from each other's potential bad behavior.

A covenant operates from a fundamentally different premise. Rather than focusing on protection from harm, it centers on the creation of shared value and meaning. A covenant says: "This is who we are together, what we stand for, how we communicate, and what we're building."

This shift from transaction to transformation, from protection to creation, represents a profound change in relationship foundation. The Covenant doesn't eliminate the need for practical agreements about responsibilities and resources, but it embeds those practical matters within a larger context of shared purpose and values.

How AI Enhances Covenant-Based Relationships

Artificial intelligence adds a powerful dimension to covenant-based relationships by providing an objective third perspective that's fully informed about your specific agreements. This addresses several challenges inherent in relationship navigation:

1. **Emotional Escalation**: During disagreements, emotions can escalate to the point where rational discussion becomes difficult. An AI system, provided with your Covenant as context, can offer perspective without the emotional charge.
2. **Blind Spots**: Every individual has areas of bias or limited perspective. AI can help identify patterns or considerations that might be in your blind spots.
3. **Creative Solutions**: When facing complex challenges, AI can generate options that neither party might have considered, expanding the solution space beyond either person's initial thinking.
4. **Consistency**: During stressful periods, people sometimes forget or drift from their core values and agreements. AI can gently remind you of the principles you've committed to when decisions are difficult.

Importantly, AI doesn't replace human discernment or decision-making. Rather, it serves as a tool that allows your best intentions—as captured in your Covenant—to remain present and accessible even during challenging moments.

Prerequisites for Successful Implementation

Before creating a Covenant of Unity, certain conditions should be in place to ensure the process succeeds:

1. **Emotional Readiness**: Both parties should approach the process from a place of relative emotional stability, not in the midst of a crisis or conflict.

2. **Genuine Voluntariness**: Each person should authentically desire the framework, not feel coerced or pressured into participation.

3. **Baseline Trust**: While the Covenant helps build deeper trust, some foundation of trust must already exist to engage meaningfully in the process.

4. **Willingness to Be Honest**: The effectiveness of your Covenant depends entirely on the honesty with which you approach its creation. Surface-level agreement without genuine commitment undermines the entire purpose.

5. **Commitment to the Process**: Creating an effective Covenant takes time and multiple conversations. Both parties should be willing to invest the necessary time for thoughtful development.

When these prerequisites are met, the creation of a Covenant becomes not just a practical exercise but a deeply connecting experience that often reveals new dimensions of understanding between participants. The process itself builds relationship strength, even before the completed document is implemented.

As we move forward to discuss the assessment phase, remember that the Covenant approach isn't meant to bureaucratize your relationships or replace spontaneity with rigid structure. Paradoxically, clear agreements create the freedom to be more fully authentic and present, because you're no longer navigating the anxiety of unstated expectations and unspoken assumptions.

Assessment: Is This Relationship Right for a Covenant?

Before embarking on the process of creating a Covenant of Unity, it's essential to assess whether a particular relationship is appropriate for this level of formal agreement. Not every connection in your life requires or would benefit from such a structured approach. This assessment phase helps you determine where to invest your relational energy and prevents the frustration of attempting to formalize relationships that aren't ready for this step.

Questions to Ask Before Creating a Covenant

Start by asking yourself these foundational questions about the relationship:

1. **Do we genuinely want to build something meaningful together?** The Covenant works best for relationships where both parties are intrinsically motivated to create something larger than themselves—whether that's a life partnership, a business venture, or a transformative friendship.

2. **Do we share fundamental values about how life should be lived?** While you don't need identical perspectives on everything, alignment in core values creates the necessary foundation for more specific agreements.

3. **Have we demonstrated reliability and follow-through with each other?** Past behavior often predicts future behavior. If there's a pattern of broken promises or inconsistency, address these issues before attempting to formalize agreements.

4. **Can we discuss difficult topics without destructive conflict?** The creation process requires candid conversations about sensitive topics. If your communication frequently breaks down into attacks, defensiveness, or withdrawal, develop better communication skills before proceeding.

5. **Are we both willing to be flexible and adapt as we grow?** The Covenant is a living document that evolves as you and your relationship evolve. Rigidity and resistance to change will undermine its effectiveness.

If you can answer these questions affirmatively, you likely have a strong foundation for developing a Covenant. If not, consider whether addressing the areas of concern might strengthen the relationship before formalizing agreements.

Evaluating Alignment in Values and Ethics

The Seven Policies explored in Chapter 4 provide an excellent framework for evaluating ethical alignment in your relationships. While perfect agreement on every aspect isn't necessary, substantial alignment in these core areas creates a solid foundation:

1. **Identity and Self-Conception**: Do you share a basic understanding of what it means to be human and how you relate to your thoughts, emotions, and body?
2. **Relationship to Truth**: Do you both value personal discernment and resist groupthink? Do you have compatible approaches to determining what's true?
3. **Communication Standards**: Do you both commit to clear definitions and precise language? Can you teach concepts to each other effectively?
4. **Moral Framework**: Do you evaluate actions similarly based on their survival value across different domains of life?
5. **Approach to Mistakes**: Do you both view errors as learning opportunities rather than moral failings? Are you willing to take responsibility without self-punishment?
6. **Economic Exchange Philosophy**: Do you share views on fairness, abundance, and mutual benefit in exchanges?

7. **Ethical Decision-Making**: Do you apply similar wisdom to choices about what you could do versus what you should do?

Discussing these areas openly can reveal whether you're sufficiently aligned to proceed with a more detailed Covenant. Significant disagreements in these fundamental areas suggest that more foundational work is needed before formalizing specific agreements.

Red Flags and Warning Signs

While the assessment process should be approached with optimism, certain patterns warrant serious caution:

1. **Resistance to Clarity**: If someone consistently avoids clear agreements or gets uncomfortable when pinning down specifics, they may not be ready for the accountability that comes with explicit commitments.
2. **History of Ethical Violations**: Past patterns of broken trust, dishonesty, or ethical compromises suggest that formal agreements may not be respected.
3. **Power Imbalance**: The Covenant works best between equals. If one person holds significantly more power (financial, emotional, or social), the process may become coercive rather than collaborative.

4. **Fundamental Value Conflicts**: Disagreements about core aspects of life—such as integrity, honesty, or responsibility—cannot be resolved through better agreements and may indicate fundamental incompatibility.
5. **Unwillingness to Evolve**: If either party views change as threatening rather than natural, the living document aspect of the Covenant will be undermined.

These warnings don't necessarily mean abandoning the relationship entirely, but they suggest that additional growth or healing may be needed before formal agreements would be effective.

The Importance of Mutual Enthusiasm

For the Covenant process to succeed, both parties must approach it with genuine enthusiasm rather than reluctant compliance. When one person pushes for formalized agreements while the other participates half-heartedly, the resulting document lacks legitimacy and rarely survives challenges.

Signs of authentic enthusiasm include:

- Asking thoughtful questions about the process
- Suggesting additional areas to address
- Referencing the future implementation with positive anticipation
- Contributing actively to the creation process

- Expressing relief or excitement about gaining clarity

If these indicators are absent, consider whether timing is the issue. Some people need time to process new approaches before fully embracing them. A gradual introduction to the concept, perhaps starting with agreements in just one area, can build confidence and demonstrate value before expanding to a comprehensive Covenant.

COVENANT OF UNION

Understanding that This Tool Reveals Incompatibilities

One of the most valuable—though sometimes painful—aspects of the Covenant process is its ability to reveal incompatibilities that might otherwise remain hidden for years. By bringing implicit expectations into explicit focus, you discover whether you're truly aligned in ways that matter for your particular relationship.

This revealing function means that not every Covenant process will result in a completed document. Sometimes, the most valuable outcome is the recognition that a particular relationship isn't suited for the level of commitment or alignment you seek. This realization, while disappointing, ultimately saves both parties from years of frustration and conflict.

Approach the assessment phase with courage and honesty. The clarity you gain—whether it confirms compatibility or reveals misalignment—serves your long-term well-being better than comfortable illusions. Remember that the goal isn't to force every relationship into a formal structure, but to identify which connections have the foundation necessary for this level of intentional co-creation.

With a clear assessment in hand, you can move forward to the practical process of creating your Covenant of Unity with confidence that you're investing in a relationship with genuine potential for harmonious growth.

Creating Your Own Covenant of Unity: A Step-by-Step Guide

The process of creating a Covenant of Unity requires thoughtfulness and care. This isn't a document to rush through in a single sitting, but rather a collaborative journey that itself strengthens your relationship. Let me guide you through each step of this process, drawing from years of experience helping different types of partnerships develop these frameworks.

Preparation: Setting the Right Environment and Mindset

Begin by creating conditions conducive to meaningful conversation:

1. **Choose the Right Timing**: Select a period when both of you are relatively free from external stressors. Avoid trying to create your Covenant during major life transitions or crises.
2. **Create a Supportive Environment**: Find a comfortable, private space where you won't be interrupted. Consider neutral territory if you're creating a Covenant with someone other than a romantic partner.
3. **Establish Ground Rules**: Agree that during Covenant discussions, you'll practice active listening, avoid interruptions, and focus on understanding rather than judgment.

4. **Approach with Curiosity**: Adopt a mindset of discovery rather than confirmation. You're exploring each other's perspectives, not simply documenting what you already think you know.

5. **Remove Distractions**: Set aside phones and other devices that might interrupt the flow of conversation. Give this process your full attention.

With these preparations in place, you're ready to begin the substantive conversations that will form your Covenant.

Initial Conversations: Discussing Hopes, Fears, and Expectations

Before drafting specific sections, engage in open-ended conversations about your relationship vision:

1. **Relationship History**: Discuss what has worked well in your relationship so far. What moments stand out as examples of the connection you want to maintain and deepen?

2. **Individual Hopes**: Share what each of you hopes to create together. What does success look like for your relationship in one year? Five years? Beyond?

3. **Personal Fears**: Vulnerably discuss what concerns you might have about the relationship. What patterns from past experiences do you want to avoid repeating?

4. **Unspoken Expectations**: Explore assumptions you've been making about how the relationship "should" function. What unstated rules have you been operating under?
5. **Growth Areas**: Identify aspects of the relationship that could be strengthened through more explicit agreements and clearer understanding.

These initial conversations create the foundation for your Covenant by revealing what matters most to each of you and where potential misalignments might exist. Take notes during these discussions—not to draft the document yet, but to capture insights you'll want to reference later.

Core Components Every Covenant Should Include

While each Covenant of Unity is unique to the specific relationship it serves, certain elements are essential for creating a comprehensive framework:

1. **Preamble**: A brief statement affirming your shared commitment and the purpose of your Covenant. This sets the tone and context for everything that follows.
2. **Shared Vision Statement**: A concise articulation of what you're creating together—the "why" behind your relationship. This might be a single compelling sentence or a short paragraph.

3. **Core Values**: The fundamental principles that guide your decisions and interactions. These typically include concepts like honesty, respect, growth, and service, defined specifically for your relationship.

4. **Principles of Partnership**: The operational guidelines for how you'll function together, covering areas like decision-making, conflict resolution, and mutual support.

5. **Roles and Responsibilities**: Clarity about what each person contributes to the relationship based on strengths, interests, and circumstances.

6. **Communication Framework**: Specific agreements about how and when you'll communicate, including regular check-ins, feedback approaches, and sensitive topic handling.

7. **Approach to Conflict**: Protocols for addressing disagreements before they escalate, including cooling-off periods, mediation options, and reconciliation practices.

8. **Review Process**: A schedule and method for regularly reviewing and updating your Covenant as your relationship evolves.

Beyond these core elements, additional sections can address specific aspects relevant to your particular relationship type, such as financial arrangements for business partners or intimate relationship boundaries for romantic partners.

Customization for Different Relationship Types

The basic Covenant framework can be adapted for various relationship contexts:

For Romantic Partnerships:

- Add sections on household management, family planning, and intimacy
- Include agreements about privacy, social media, and extended family relationships
- Address financial integration and long-term planning

For Business Partnerships:

- Expand on decision authority and approval processes
- Include resource allocation and profit distribution
- Address client/customer relationship management
- Detail growth strategy alignment and exit planning

For Close Friendships:

- Focus on support expectations and boundaries
- Address confidentiality and trust parameters
- Include agreements about involving other friends and handling conflicts of loyalty

For Family Relationships:

- Emphasize respect for individual autonomy while maintaining connection
- Address historical patterns and trigger management
- Include agreements about holidays, traditions, and conflict management

In each case, the customization process involves identifying the most common sources of friction in that particular relationship type and creating specific agreements to address them proactively.

The Drafting Process

Once you've had thorough initial conversations, proceed to drafting your Covenant:

1. **Start with Templates**: Use the blank Covenant of Unity template as a starting point, recognizing that you'll modify it significantly to fit your specific relationship.
2. **Draft Separately, Then Compare**: Have each person independently draft their ideal version of certain sections, then compare to identify areas of alignment and difference.
3. **Prioritize Sections**: Don't try to complete the entire document at once. Begin with the most important sections for your particular relationship, then build from there.

4. **Use Clear, Specific Language**: Avoid vague terms like "regular communication" or "fair contribution." Instead, specify "daily check-ins of at least 15 minutes" or "financial contributions proportional to income with a minimum of X."

5. **Document Areas of Uncertainty**: When you're not ready to make definitive agreements in certain areas, acknowledge this explicitly rather than creating false consensus.

6. **Incorporate Flexibility Mechanisms**: Include processes for handling unexpected situations and adapting agreements as needs change.

The drafting process typically requires multiple sessions spread over several weeks. This timeline isn't about bureaucratic delay but about allowing insights to emerge between sessions and ensuring the document truly reflects your considered intentions rather than hasty compromises.

Review and Refinement

Once you have a complete draft, engage in a thorough review process:

1. **Individual Reflection**: Take time for each person to review the draft independently, noting areas that feel particularly aligned or potentially problematic.

2. **Comprehensive Discussion**: Meet to discuss the document section by section, ensuring that each agreement genuinely reflects mutual understanding and commitment.
3. **Living Document Mentality**: Remind yourselves that this isn't a final, unchangeable contract but a current best understanding that will evolve as you both grow.
4. **Deal-Breaker Identification**: Honestly discuss whether any elements feel impossible to commit to, recognizing that these areas require either renegotiation or acceptance of difference.
5. **Practical Testing**: Consider "trying out" certain agreements for a short period before finalizing them, especially for aspects that represent significant changes to your current patterns.

This review process ensures that your Covenant represents genuine consensus rather than surface agreement that masks underlying concerns.

Formal Adoption and Celebration

The completion of your Covenant deserves recognition and ceremonial acknowledgment:

1. **Formal Signing**: Set aside a special time to sign the document together, marking your commitment to its principles.

2. **Celebration Ritual**: Create a meaningful ritual that symbolizes your shared journey—perhaps a special meal, exchange of meaningful objects, or visit to a significant location.

3. **Commemoration**: Consider creating a physical representation of your Covenant beyond the document itself—a piece of art, a shared symbol, or another tangible reminder of your commitments.

This ceremonial aspect shouldn't be dismissed as mere sentimentality. It helps anchor your agreements in emotional memory and creates a milestone that marks the thoughtful intention behind your relationship.

Remember that creating your Covenant is not the end but the beginning of a new phase in your relationship—one characterized by greater clarity, intentionality, and shared purpose. The document itself matters less than the conversations it generates and the aligned understanding it creates. Approach the process with patience, curiosity, and genuine care for creating something beautiful together.

Essential Elements of the Covenant of Unity

Now that we've covered the process for creating your Covenant, let's explore each essential element in greater detail. Understanding the purpose and structure of these components will help you develop a comprehensive document that addresses the unique needs of your relationship.

Shared Vision Statement

The vision statement serves as the North Star for your relationship—a compelling description of what you're creating together that inspires and guides all other agreements. This isn't merely a list of goals but a vivid articulation of the relationship's purpose and potential.

An effective vision statement:

- Speaks to both hearts and minds
- Extends beyond immediate circumstances to long-term aspiration
- Reflects values that both parties genuinely share
- Creates excitement and energy when read
- Provides direction during challenging times

For example, a business partnership might declare: "Together, we're building a company that transforms healthcare accessibility while creating fulfilling work for our team and sustainable value for our communities."

A romantic partnership might envision: "We are creating a relationship that serves as a sanctuary of love, growth, and joy for ourselves and a model of harmony for those around us."

The specifics matter less than the resonance—your vision statement should evoke a sense of "Yes, that's exactly what we want to build together" from both parties.

Core Values

Values provide the foundation for all decisions and actions within your relationship. While people often assume they share values without discussion, deeper exploration frequently reveals different interpretations or priorities. Your Covenant should explicitly define 4-7 core values that guide your partnership.

For each value, include:

- A clear name (e.g., "Integrity," "Innovation," "Compassion")
- A brief definition that captures your shared understanding
- How this value manifests in practical behavior

This clarity prevents the common scenario where both people claim to value the same thing but mean fundamentally different things by it. For instance, both might value "respect," but one person defines it as giving space and privacy while the other defines it as regular check-ins and involvement.

Principles of Partnership

While values define what you stand for, principles describe how you operate together. These are the fundamental agreements that guide your interactions across all contexts.

Essential principles to define include:

- **Decision-Making**: Will you make all significant decisions by consensus? Are certain domains delegated to one partner? What constitutes a "significant" decision requiring joint input?
- **Truth-Telling**: How will you ensure honesty while maintaining kindness? What commitments do you make about sharing difficult truths?
- **Mutual Support**: How will you balance independence and interdependence? What can each person expect in terms of emotional, practical, and other forms of support?
- **Growth Orientation**: How will you challenge each other to evolve while maintaining acceptance? What commitments do you make to supporting each other's individual development?

These principles create the operational framework for your relationship, ensuring that day-to-day interactions align with your larger vision and values.

Spiritual Practices (If Applicable)

For many relationships, spiritual alignment creates profound connection and meaning. If this dimension is relevant to your partnership, include agreements about:

- Shared practices that deepen your connection to each other and to larger meaning
- Respect for differences in spiritual expression or experience
- How spiritual values inform practical decisions
- Ways you'll support each other's spiritual growth

This section isn't only for traditionally religious partnerships—it applies to any relationship that acknowledges dimensions beyond the material and practical.

Roles and Responsibilities

Clarity about who does what prevents the resentment that often builds when expectations go unstated. Your Covenant should outline:

- Each person's areas of strength and primary contributions
- Specific responsibilities each person commits to fulfilling
- How tasks and responsibilities will be reviewed and potentially redistributed over time
- Processes for handling situations when someone can't fulfill their usual responsibilities

This clarity doesn't create rigid boxes but rather establishes a baseline understanding that can evolve as circumstances change.

Communication Framework

Communication is so fundamental to relationship success that it deserves detailed attention in your Covenant. Include agreements about:

- Regular check-ins and their format (daily connections, weekly reviews, monthly deeper conversations)
- How you'll share feedback with each other
- Approaches for discussing sensitive topics
- Commitments about digital communication (response times, appropriate channels for different topics)
- Privacy boundaries regarding what can be shared with others

These agreements create reliable connection patterns that maintain relationship strength even during busy or challenging periods.

Rituals and Practices

Relationships thrive on consistent practices that reinforce connection and shared identity. Your Covenant should include rituals that:

- Celebrate successes and milestones
- Maintain connection during ordinary times

- Repair connection after conflicts or distance
- Mark transitions and new phases

These might include daily appreciation practices, monthly date nights, annual retreats to review your Covenant, or other regular activities that strengthen your bond.

Approach to Conflict and Challenges

Every relationship faces conflicts and challenges. Your Covenant should outline how you'll navigate these inevitable difficulties:

- Early warning systems for identifying potential conflicts
- Protocols for addressing disagreements before they escalate
- Communication guidelines during emotional tension
- When and how to involve third parties for perspective or mediation
- Specific steps for repair and reconciliation after conflicts

Having these agreements in place before conflict arises allows you to handle difficulties as a team rather than as adversaries.

Ethical Commitments

Relationships exist within larger contexts of ethical responsibility. Your Covenant should address:

- Commitments to honesty and integrity within and beyond the relationship
- How you'll handle situations where values seem to conflict
- Approaches to creating positive impact beyond your relationship
- Ways you'll hold each other accountable to your highest standards

These commitments ensure your relationship contributes positively to the world rather than becoming self-enclosed or ethically compromised.

Vision for the Future

While the vision statement captures your overall purpose, this section addresses more specific long-term aspirations:

- Key milestones you hope to achieve together
- How you envision the relationship evolving over time
- Legacy considerations—what you hope to create that outlasts the relationship itself
- Adaptability principles for navigating major life changes

This forward-looking perspective helps maintain momentum and shared purpose through the inevitable changes life brings.

Shared Mission Projects

Many of the most fulfilling relationships involve creating something together beyond the relationship itself. Include:

- Specific initiatives or projects you commit to co-creating
- Each person's contributions to these shared missions
- Timelines and milestones for implementation
- How you'll evaluate success and make adjustments

These shared projects create opportunities for collaboration, growth, and meaningful impact that strengthen your connection.

Review and Adaptation Process

Finally, your Covenant should include specific plans for keeping it relevant and aligned with your evolving relationship:

- Scheduled times for reviewing the entire document (annually or at major transitions)
- Processes for proposing and implementing changes
- How you'll track progress on commitments and goals
- Celebrations for milestones and renewal points

This process ensures your Covenant remains a living document that grows with your relationship rather than becoming an outdated artifact.

By thoughtfully addressing each of these elements, you create a comprehensive framework that supports your relationship across all dimensions. Remember that the goal isn't perfection or rigid adherence to a formula—it's creating sufficient clarity and alignment to allow your unique relationship to flourish with minimal friction and maximum joy.

The Security Dimension: Protecting Your Relationship in the Digital Age

In an era of advanced AI, deepfakes, and sophisticated digital manipulation, relationships face security challenges that previous generations never encountered. Creating protocols to verify identity and maintain security has become an essential aspect of relationship agreements, particularly for partnerships with significant shared responsibilities or resources.

Creating Unique Verification Systems

Every close relationship benefits from establishing private verification methods that only the involved parties know. Consider implementing these approaches:

1. **Personal Authentication Questions**: Develop a set of questions with answers only you and your partner would know. These should go beyond basic information (like "Where did we meet?") to include nuanced details from shared experiences.

2. **Code Words and Phrases**: Establish specific words or phrases that signal authentic communication. These can be particularly valuable in text conversations where voice and visual cues are absent.

3. **Sequential Verification**: Create multi-step verification protocols for sensitive situations. For example, a business partner might request both a code word and a specific personal reference before proceeding with a major financial transaction discussed via email.

What makes these systems effective is their personalization—they should incorporate details from your unique relationship history that would be extremely difficult for an outsider to replicate, even with advanced AI tools.

Protecting Against Digital Impersonation

Deepfake technology makes voice and video impersonation increasingly convincing. Protect your relationship with these practices:

1. **Contextual Verification**: Establish the habit of referencing recent shared experiences or ongoing conversations when making unusual requests. This creates a continuity that's difficult for impersonators to maintain.

2. **Secondary Channel Confirmation**: For significant decisions or requests, confirm through a different communication channel. If you receive an unusual text request, follow up with a video call before taking action.

3. **Predetermined Security Levels**: Create a system of security levels for different types of requests or decisions, with corresponding verification requirements. Minor changes might need minimal verification, while major decisions require multiple confirmation steps.

These practices balance security with practical usability, recognizing that different situations require different levels of verification.

Emergency Protocols

Every relationship should have clear agreements about handling emergency situations:

1. **Duress Signals**: Establish subtle indicators that communicate you're under pressure or not free to speak openly. These could be specific phrases, word orders, or references that wouldn't arouse suspicion from others but would alert your partner.

2. **Emergency Contacts and Procedures**: Document who to contact and what steps to take in various emergency scenarios. This might include medical emergencies, security threats, or situations where one partner is unreachable.

3. **Recovery Procedures**: Create protocols for re-establishing secure communication if you suspect your usual channels have been compromised.

These emergency protocols provide peace of mind and practical guidance during high-stress situations when clear thinking might be compromised.

Digital Security Practices

Beyond interpersonal verification, implement technical security measures to protect your shared information:

1. **Shared Password Management**: Use secure password managers to share necessary credentials without compromising security. Establish clear agreements about which accounts are individual versus shared.

2. **Encryption for Sensitive Communications**: Utilize encrypted messaging platforms for sensitive discussions, especially regarding finances, legal matters, or private personal information.

3. **Regular Security Audits**: Schedule periodic reviews of your digital security practices, updating them as technologies and threats evolve.

These technical practices complement your interpersonal verification systems, creating layered protection for your relationship.

Balancing Security with Trust and Convenience

While security measures are important, they must be balanced with practical usability and the fundamental trust that defines your relationship:

1. **Appropriate Security Scaling**: Match security measures to actual risk levels. High-profile business partnerships might require more extensive verification than casual friendships.

2. **Regular Reassessment**: Periodically evaluate whether your security protocols still serve your relationship or have become unnecessary burdens.

3. **Trust as Foundation**: Remember that security measures supplement trust rather than replace it. The goal is protection against external threats, not creating suspicion between partners.

The most effective security systems are those that become natural habits, providing protection without creating constant friction or paranoia.

By thoughtfully incorporating these security dimensions into your relationship agreements, you create resilience against modern threats while maintaining the ease and spontaneity that allow relationships to flourish. In a world where digital impersonation becomes increasingly sophisticated, these shared protocols serve as a protective boundary around your relationship, ensuring that your connection remains authentic and secure.

Implementing the Covenant in Daily Life

Creating a Covenant of Unity represents an important beginning, but its true value emerges through consistent implementation in your daily interactions. This section explores how to move from document to practice, integrating your agreements into the fabric of your relationship.

Regular Review Sessions

The Covenant is designed to be a living document, not a static artifact. Establish a rhythm of regular reviews to maintain its relevance:

1. **Weekly Check-ins**: Dedicate 15-30 minutes each week to briefly discuss how well you're adhering to your communication agreements and addressing any minor adjustments needed.
2. **Monthly Deep Dives**: Set aside 1-2 hours monthly to explore one section of your Covenant in detail, discussing what's working well and what might need refinement.

3. **Quarterly Assessments**: Conduct more comprehensive reviews quarterly, evaluating progress on shared projects and how well your agreements are supporting your vision.

4. **Annual Renewal**: Schedule a more ceremonial annual review where you revisit your entire Covenant, celebrating accomplishments, addressing challenges, and recommitting to your shared vision.

These regular touchpoints prevent the common pattern where agreements slowly fade from active consciousness until they're rediscovered during conflicts—precisely when they're most difficult to implement.

Using the Covenant as a Decision-Making Framework

When facing choices or challenges, actively reference your Covenant rather than relying on memory or impulse:

1. **Values-Based Decisions**: When evaluating options, explicitly discuss how each alternative aligns with your core values as defined in your Covenant.

2. **Role Clarity**: Use your roles and responsibilities section to determine who should take the lead on particular decisions or tasks.

3. **Conflict Prevention**: Before disagreements escalate, refer to your conflict resolution protocols, following the

steps you outlined when you were in a more objective state of mind.

4. **Opportunity Evaluation**: When new opportunities arise, assess them against your shared vision and long-term goals to determine whether they merit your investment of time and energy.

This conscious application transforms your Covenant from a theoretical document into a practical tool for navigating daily life together.

Leveraging AI to Support Problem-Solving

Artificial intelligence offers powerful capabilities for implementing your Covenant effectively:

1. **Context-Aware Guidance**: By providing your AI system with your Covenant as context, you enable more personalized recommendations during challenging situations. The AI can remind you of specific agreements relevant to the current issue without the emotional charge that sometimes accompanies human reminders.

2. **Pattern Recognition**: AI can help identify recurring patterns in your relationship that might warrant attention—perhaps certain types of decisions consistently create tension, or specific communication styles repeatedly lead to misunderstanding.

3. **Resource Suggestions**: When facing challenges in implementing specific aspects of your Covenant, AI can suggest relevant resources, exercises, or approaches tailored to your particular situation.

4. **Objective Perspective**: During disagreements, consulting an AI that has your Covenant as context can provide a neutral third perspective focused on your shared commitments rather than either individual's position.

This integration of AI support works best when both parties have agreed to it as part of your implementation approach, with clear boundaries about what information is shared and how the guidance will be used.

Evolving the Document as the Relationship Grows

The Covenant should evolve as your relationship and circumstances change:

1. **Formal Amendments**: Create a simple process for proposing and adopting amendments to your Covenant as new insights emerge or circumstances change.

2. **Living Document Technology**: Consider using digital platforms that support collaborative editing and version control, allowing your Covenant to evolve while maintaining a history of your agreements.

3. **Growth Markers**: Use changes to your Covenant as opportunities to recognize and celebrate how your relationship has matured and developed over time.
4. **External Facilitation**: For significant revisions, consider engaging a trusted third party (human or AI) to facilitate the conversation, ensuring all perspectives are heard and integrated.

This evolution ensures your Covenant remains relevant rather than becoming an outdated artifact that no longer reflects your current relationship reality.

Handling Breaches or Misalignments

Even with the best intentions, there will be times when actions don't align with agreements. Establish clear processes for addressing these situations:

1. **Compassionate Accountability**: When agreements aren't honored, approach the situation with curiosity rather than accusation. Most breaches stem from misunderstanding, forgetfulness, or competing priorities rather than malicious intent.
2. **Repair Protocols**: Develop specific steps for repairing trust and realigning with agreements after breaches occur. These might include acknowledgment, understanding the impact, recommitment, and concrete action to prevent recurrence.

3. **Learning Orientation**: Treat misalignments as valuable data about where your Covenant might need clarification or where implementation support is needed, rather than as evidence of failure.

4. **Proportional Response**: Match your response to the significance of the breach. Minor deviations require simple course correction, while fundamental violations may necessitate deeper conversation about commitment to the relationship itself.

This approach to handling inevitable imperfections keeps your implementation realistic and sustainable rather than becoming a source of shame or rigidity.

By thoughtfully implementing your Covenant in these ways, you transform it from a theoretical exercise into a practical foundation for relationship flourishing. The goal isn't perfect adherence to every detail but rather the creation of a shared framework that continuously guides you toward greater alignment, mutual understanding, and collective purpose.

Remember that implementation is itself a skill that develops with practice. Begin with the elements most crucial to your particular relationship, gradually incorporating additional aspects as your implementation capacity grows. Over time, many of these practices will become natural habits that support your relationship with minimal conscious effort.

Troubleshooting and Advanced Implementation

Even with careful preparation and thoughtful design, you'll encounter challenges when implementing your Covenant of Unity. This section addresses common obstacles and provides advanced strategies to refine your approach.

When Enthusiasm Is Unbalanced

One of the most common challenges occurs when one partner is more invested in the Covenant process than the other. If you find yourself in this situation:

1. **Start Small**: Rather than pushing for a comprehensive Covenant immediately, begin with agreements in just one area that both parties genuinely want to improve.

2. **Focus on Benefits**: Emphasize specific ways the framework will address challenges the less enthusiastic partner has identified, making the process relevant to their concerns.

3. **Allow for Personalization**: Invite the hesitant partner to suggest modifications to the process that would make it feel more comfortable and authentic to them.

4. **Respect Timing**: Recognize that some people need time to process new approaches. Introducing the concept, then

revisiting it after a period of reflection, often yields greater engagement.

The key is patience—forced agreement creates superficial compliance rather than genuine commitment. A partial Covenant built on mutual enthusiasm serves better than a comprehensive document where commitment is imbalanced.

COVENANT
OF UNITY

Adapting for Cultural Differences

When partners come from different cultural backgrounds, additional considerations help ensure your Covenant respects diverse perspectives:

1. **Explicit Value Translation**: Discuss how similar values might be expressed differently across cultural contexts. For example, "respect" might manifest as direct communication in one culture and tactful indirectness in another.
2. **Decision Hierarchy Awareness**: Acknowledge cultural differences in how decisions typically involve family or community members, adapting your frameworks accordingly.
3. **Ritual Integration**: Incorporate meaningful cultural practices from both backgrounds into your relationship rituals and celebration approaches.
4. **Terminology Sensitivity**: Choose language that resonates across cultural contexts, avoiding terms that carry negative connotations in either partner's background.

These adaptations honor cultural diversity while creating shared understanding that transcends cultural differences.

Integrating with Legal Agreements

In some relationships, particularly business partnerships or marriages, your Covenant exists alongside formal legal agreements. Navigate this relationship by:

1. **Clarifying Domains**: Distinguish which aspects of your relationship are governed by legal contracts versus your Covenant, understanding their different purposes.
2. **Alignment Assessment**: Verify that your Covenant commitments don't contradict legal obligations, consulting professional advisors when necessary.
3. **Spirit vs. Letter**: Use your Covenant to articulate the spirit and intentions behind legal agreements, creating context that guides implementation.
4. **Complete Coverage**: Ensure your Covenant addresses relationship dimensions beyond legal considerations, particularly emotional, communication, and growth aspects.

This integration allows your relationship to benefit from both the protection legal agreements provide and the relational depth the Covenant fosters.

Using the Covenant as a Diagnostic Tool

Beyond its primary function as a guiding framework, your Covenant can serve as a powerful diagnostic tool for relationship health:

1. **Pattern Identification**: Notice which sections of your Covenant require frequent discussion or renegotiation, revealing areas that may need deeper exploration.
2. **Value Drift Detection**: Regular reviews can highlight when individual values have evolved in ways that create potential misalignment, allowing proactive adjustment.
3. **Energy Assessment**: Pay attention to which aspects of your Covenant generate enthusiasm versus resistance during reviews, indicating shifts in priorities or interests.
4. **Implementation Gaps**: Consistent difficulty implementing specific agreements often signals underlying issues that require attention beyond the agreement itself.

This diagnostic function helps you address emerging challenges before they become relationship-threatening problems.

When to Revise Versus When to Walk Away

Perhaps the most challenging aspect of Covenant implementation is discerning when to revise agreements versus when to recognize fundamental incompatibility:

1. **Healthy Revision Indicators**:
    - Both parties remain committed to the shared vision
    - Changes reflect growth rather than compromise of core values

- Adjustments address implementation challenges rather than fundamental principles
- The revision process feels energizing rather than depleting

2. **Potential Incompatibility Signals**:
   - Persistent pattern of breached agreements despite good-faith efforts
   - Fundamental value conflicts that create ongoing tension
   - One or both parties feel consistently diminished by the relationship
   - Revision discussions repeatedly generate conflict rather than clarity

The Covenant exists to serve the relationship, not to perpetuate connections that have become fundamentally misaligned. Sometimes the most appropriate outcome of your Covenant process is the recognition that a particular relationship has served its purpose and reached its natural conclusion.

Having the courage to acknowledge this reality—with the clarity and compassion your Covenant process has fostered—allows both parties to move forward with integrity rather than remaining in patterns that no longer serve their highest good.

Conclusion: The Transformative Power of Intentional Agreements

As we conclude this exploration of relationship agreements, I invite you to consider a fundamental paradox: structure creates freedom. This principle, counterintuitive at first glance, reveals itself clearly through the Covenant framework.

When relationships operate with implicit expectations and unspoken rules, anxiety inevitably arises. You move cautiously, uncertain whether your actions will violate invisible boundaries or unspoken expectations. This caution constrains authentic expression and spontaneity.

In contrast, clear agreements create a defined space within which you can move freely. When you know exactly where the boundaries lie, what's expected in specific situations, and how to address inevitable challenges, you're liberated to be fully present and authentic within those parameters.

This structured freedom extends beyond practical matters to the emotional dimension. When agreements clarify how feedback will be shared, conflicts addressed, and differences honored, emotional safety emerges. This safety creates the conditions for vulnerability, creativity, and genuine connection that superficial harmony never achieves.

The Covenant of Unity represents a comprehensive approach to creating this structured freedom. By addressing the full spectrum of relationship dimensions—from practical responsibilities to spiritual connection, from communication practices to conflict resolution—it provides a foundation for relationships that don't just survive but thrive amid life's inevitable changes and challenges.

As you consider implementing this approach in your own relationships, remember that the goal isn't bureaucratic perfection or rigid adherence to rules. The goal is creating sufficient clarity and alignment that your energy can focus on what truly matters: growing together, supporting each other's development, and contributing to something larger than yourselves.

The process itself—the thoughtful conversations, honest self-reflection, and mutual commitment to creating something beautiful together—often proves as valuable as the resulting document. These conversations build understanding, empathy, and connection that transform relationships independently of the specific agreements they produce.

I encourage you to begin where you are, with the relationships that matter most in your life. Start with one section of the Covenant that seems most relevant to your current situation. Experience how the process of creating explicit agreements transforms your understanding and connection. Then expand gradually, building a comprehensive framework that evolves with your relationship.

The journey of intentional relationship design isn't always easy. It requires courage to examine assumptions, honesty to acknowledge expectations, and vulnerability to express needs and boundaries. Yet the rewards—relationships characterized by clarity, purpose, and genuine connection—far outweigh the challenges of the process.

In a world where relationships often drift into dysfunction through negligence rather than intention, the conscious choice to create clear agreements represents a revolutionary act. It declares that your relationships are too important to leave to chance or implicit understanding—that they deserve the same thoughtful design you bring to other significant aspects of your life.

I invite you to embrace this revolution in your own relationships, creating agreements that transform connection from a source of stress and uncertainty into a foundation of joy, growth, and meaningful contribution to the world you share.

Part IV

CALL TO ACTION

Chapter 9: Creating Harmonious Communities

Introduction: Making a Difference in Your Community

In your heart, don't you long to create something beautiful in the world around you? This desire to make a difference burns within most of us—a quiet but persistent flame that seeks expression through service to our neighborhoods, faith communities, cities, and beyond.

For generations, we've channeled this longing through traditional paths: volunteering at local organizations, campaigning for public office, or creating businesses that elevate our communities. These efforts, while noble, have always been constrained by the limits of our individual knowledge, the expertise immediately available to us, and the resources we could personally gather.

Today, everything has changed.

Artificial intelligence has shattered these constraints. The methodology I've shared throughout this book—the triangle of **Ethics**, **Communication**, and **Context**—now gives you unprecedented power to transform the places and communities you care about most. You can access the collective wisdom of successful community projects worldwide, implement sophisticated frameworks without specialized training, and create solutions that would have seemed like fantasy just a few years ago.

This revolution arrives at the perfect moment. Many community organizations continue struggling with outdated tools and limited resources while facing increasingly complex challenges. Environmental degradation, social inequality, educational disparities, healthcare access—these issues demand innovation at a scale traditional approaches simply cannot provide.

In this final chapter, we'll explore how to channel this new power toward creating the world we wish to see. Whether you're a seasoned community leader or just beginning your journey of service, what follows will help you create lasting, meaningful change in the places that matter most to you.

The Project Management Approach to Community Transformation

At its core, every community initiative—from neighborhood clean-ups to major infrastructure developments—is a project. This simple insight allows us to apply proven management principles to our community work, dramatically increasing our chances of success.

The Project Management Institute offers a powerful framework called the **triple constraint** that gives us a clear lens through which to view any community initiative:

- **Project Scope**: What exactly are we creating? What falls within our boundaries, and what doesn't? What specific outcomes will signal success?
- **Project Schedule**: When does our work begin and end? What key milestones must we achieve along the way? In what sequence should activities unfold?
- **Resources**: What people, funds, equipment, materials, and knowledge do we need to accomplish our goals?

SCOPE
RESOURCES
SCHEDULE

Historically, resources have been the greatest constraint on community projects. Even the most passionate leaders and volunteers have struggled against the limits of their own knowledge and access to expertise. The internet expanded our reach somewhat, but transformative information remained scattered, unorganized, and often inaccessible to those who needed it most.

Artificial intelligence fundamentally changes this equation. Through large language models and The Hasan Method™, any community leader can now access the practical wisdom of virtually every successful similar initiative ever documented. This democratization of expertise represents the most significant advancement in community development in our lifetime.

Think of the typical challenges facing community projects:

- How do we design this initiative for maximum impact?
- Who should we involve, and how do we coordinate their efforts?
- What pitfalls have similar projects encountered, and how can we avoid them?
- Where might we find funding and other resources?
- How do we measure success and adapt our approach as we learn?

For each of these questions, AI offers not just answers but wisdom—not just information but contextual understanding drawn from thousands of similar situations. It combines the insights of experienced community leaders, project managers, subject matter experts, and implementers—all accessible through natural conversation, with no technical expertise required.

This doesn't mean AI replaces human judgment or compassion. The heart of community work remains deeply human—the desire to serve, the creativity to solve problems, the empathy to understand needs, and the persistence to overcome obstacles. AI simply amplifies these human qualities by removing the knowledge barriers that have traditionally limited their expression.

Practical Applications: Community Projects Enhanced by AI

Let me share specific examples of how AI transforms community initiatives across different domains, starting with one of the most universal challenges facing communities worldwide.

Environmental Initiatives

Have you noticed how trash accumulates in public spaces despite our best intentions? Traditional approaches often involve organizing occasional volunteer clean-ups with limited planning and resources. But with AI-enhanced project management, you can create something far more effective.

Using this methodology, you might create a markdown file specifically for trash reduction that includes:

- Detailed analysis of the most successful clean-up methodologies worldwide
- Specific strategies tailored to your community's unique challenges
- Guidance on recruiting and managing volunteers for sustained engagement
- Information about recycling programs with proven track records
- Potential partnerships with local businesses, schools, and government agencies
- Step-by-step implementation plans with clear timelines and responsibilities

Most importantly, AI helps you design not just an event but a *system*—a sustainable approach that addresses root causes and creates lasting change. This might include innovative public education campaigns, business partnerships that reduce waste generation, coordination with municipal services, and ongoing monitoring systems.

One particularly valuable application involves securing funding. Many environmental initiatives struggle due to limited resources, yet numerous grants go unclaimed because organizations don't complete applications correctly or on time. AI can transform this process, helping you identify appropriate funding sources and complete applications with precisely the information needed to succeed.

Community Revitalization

Think about the heart of your community—the main street, town center, or neighborhood hub. These spaces often need revitalization to remain vibrant, but such projects involve complex coordination among business owners, residents, government officials, and other stakeholders.

Consider the example of Laurel, Maryland, a growing city that sits within four different counties. Like many municipalities, Laurel is undertaking a main street revitalization project—a multifaceted initiative requiring careful planning and execution.

Rather than relying solely on their team's direct experience, Laurel's leaders now leverage insights from revitalization projects worldwide. This expanded knowledge helps them address every aspect of their initiative:

- Creating optimal traffic and parking solutions
- Coordinating business hours for maximum community benefit
- Designing outdoor spaces that encourage gathering and connection
- Developing programming that draws people downtown

- Implementing safety measures that make everyone feel welcome
- Preserving historical character while embracing modern needs
- Building economic development strategies for long-term sustainability

The AI-enhanced approach doesn't replace the creativity of urban planners or the practical knowledge of business owners. Instead, it helps these stakeholders access relevant case studies and innovative approaches from similar communities globally. It allows them to learn from both successes and failures elsewhere, avoiding common pitfalls while implementing proven strategies.

This application extends beyond main street projects to any community development initiative: park creation, affordable housing, transportation improvements, or recreation facilities. In each case, AI helps leaders access specialized knowledge that would otherwise remain unavailable or prohibitively expensive.

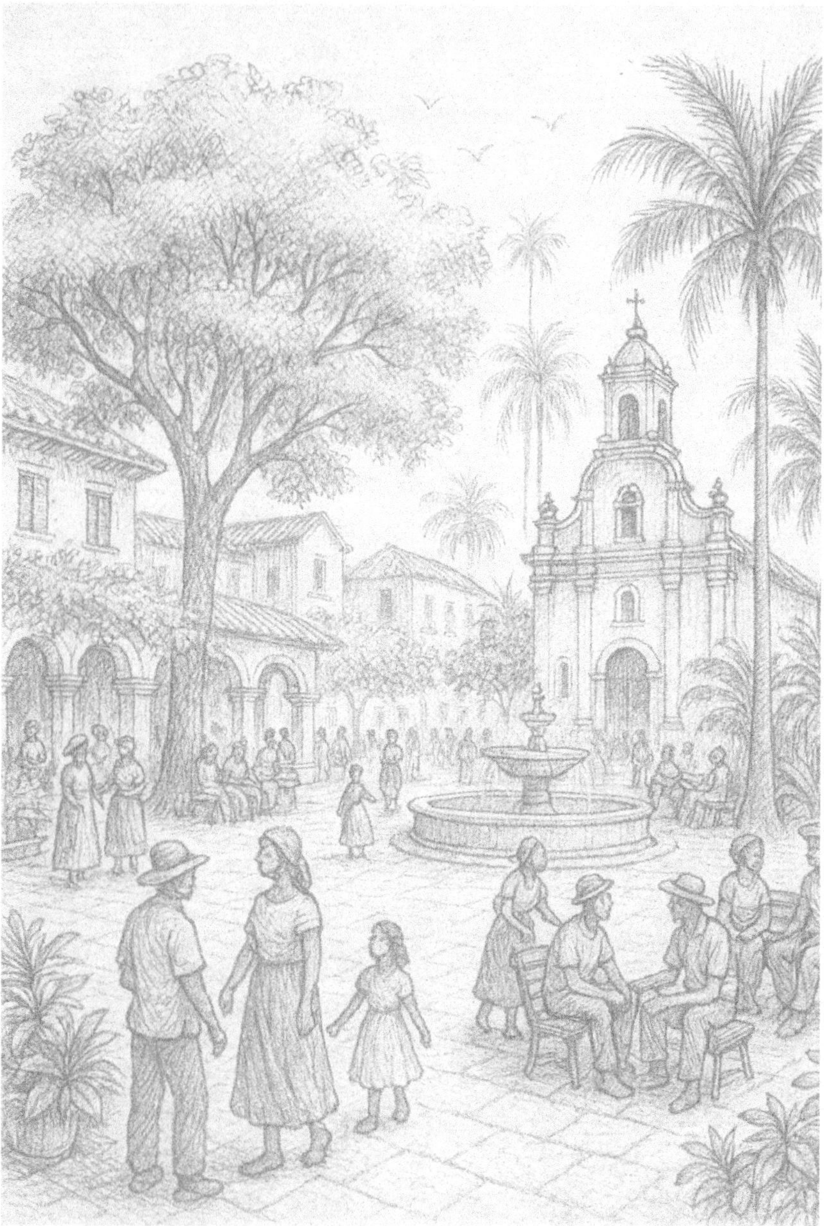

Clubs, Organizations, and Social Groups

The fabric of community life consists largely of the hobby groups, sports teams, faith communities, and social clubs where people gather around shared interests and values. These organizations, often run by volunteers with limited time and resources, stand to benefit tremendously from AI enhancement.

For any group you belong to, consider how AI might help:

- Streamline organizational processes that currently consume valuable volunteer time
- Create more engaging events based on successful models elsewhere
- Develop better communication systems that keep members connected
- Enhance recruitment and retention through proven approaches
- Secure resources through grants, partnerships, or innovative funding models
- Implement best practices from similar organizations worldwide

By creating a markdown file specific to your organization, you provide AI with the context needed to offer highly relevant, customized guidance. This might include everything from meeting formats and conflict resolution approaches to fundraising strategies and leadership development programs.

Family Planning and Event Organization

The principles of AI-enhanced management extend even to family life and personal events. Whether planning a wedding, reunion, holiday celebration, or educational opportunity for your children, the same methodology applies.

For example, if you're planning a wedding, a specialized markdown file could help organize every aspect:

- Venue selection based on your specific priorities and constraints
- Vendor coordination that ensures seamless collaboration
- Budget management that prevents unexpected costs
- Guest experience designed around your unique vision
- Risk mitigation for weather or other contingencies
- Cultural traditions and personalization opportunities

The AI doesn't make decisions for you—it expands your awareness of options, highlights potential challenges, and suggests creative possibilities you might not otherwise consider. It helps ensure you've accounted for all relevant factors while preserving the deeply personal nature of these meaningful life events.

Strategies for AI-Enhanced Project Success

To maximize impact and ensure successful outcomes when applying artificial intelligence to community initiatives, consider these specific strategies.

Exercise Persistence with AI Outputs

When working with human collaborators, we often accept "good enough" outputs to respect their time and maintain relationships. With AI, a different approach yields far better results: persistent refinement.

When an AI provides a good response, consider pushing further: "Thank you for this excellent work. How could we make it even better? What perspectives might we be missing? What alternatives should we consider?"

This persistence extracts maximum value from each interaction, refining outputs until they achieve the highest possible quality. Unlike human collaborators, the AI won't grow tired, frustrated, or offended by requests for improvement—it's designed for iterative collaboration.

I've found this approach particularly valuable when developing grant applications, strategic plans, and communication materials. What begins as a good draft often evolves into something exceptional through three or four rounds of thoughtful refinement—a process that might take weeks with human collaborators but can happen in minutes with AI.

Implement Comprehensive Risk Mitigation

For any community project, dedicated risk analysis represents perhaps the most valuable AI application. By systematically identifying potential challenges and failure points, you gain perspectives that might otherwise remain hidden until problems emerge.

For example, in a community festival, risk mitigation might address:

- Weather contingencies that could affect outdoor activities
- Security protocols for ensuring participant safety
- Accessibility considerations for people with various needs
- Vendor reliability and backup options
- Transportation and parking challenges
- Financial risks and budget safeguards

The key advantage of AI-assisted risk mitigation is its comprehensive nature. Rather than being limited by your team's particular experiences, you access a global database of things that have gone wrong in similar projects—allowing you to prepare for contingencies you might never have anticipated.

This approach doesn't eliminate all problems, but it dramatically reduces their frequency and severity. More importantly, it helps you develop response plans before crises occur, allowing for calm, effective action rather than reactive scrambling.

Create Collaborative Markdown Files

For community projects involving multiple stakeholders, consider creating centrally updated collaborative markdown files that serve as both knowledge repositories and coordination mechanisms.

These files might include:

- Vision, objectives, and success criteria that keep everyone aligned
- Stakeholder roles and responsibilities clearly defined
- Timeline and milestone tracking to ensure progress
- Resource allocation and management protocols
- Decision-making frameworks that prevent bottlenecks
- Communication standards that keep information flowing
- Best practices and lessons learned as the project progresses

By maintaining these documents collectively, you create a shared brain trust that preserves institutional knowledge even as individual participants may change over time. This approach proves particularly valuable for long-term community initiatives that span months or years and involve diverse stakeholders.

Achieve More with the Same Effort

Perhaps the most important strategy involves focusing AI not on doing more things, but on doing the same things more effectively. I'm not suggesting you take on countless additional projects or work longer hours. Instead, I'm proposing that you enhance the impact of efforts you're already making.

For those already engaged in community work—those already committed to making a difference—AI offers the opportunity to achieve dramatically better outcomes from the same investment of time, energy, and resources. This multiplication effect allows dedicated community leaders to create change at scales previously unimaginable.

Consider a small environmental organization I worked with recently. With the same staff, the same budget, and the same number of volunteer hours, they tripled their impact simply by implementing AI-enhanced planning, coordination, and resource allocation. The difference wasn't in how hard they worked but in how effectively they channeled their existing passion and commitment.

The Responsibility of Leadership in the AI Era

As we embrace these new capabilities, we must also consider the expanded responsibility they bring. The ability to create greater impact demands deeper reflection on the purpose and direction of that impact.

Moving Beyond Survival to Purpose

For those fortunate enough to have moved beyond immediate survival concerns, there exists the privilege of pursuing purpose—of examining what contribution you wish to make to the world beyond meeting your own basic needs.

When we operate in survival mode, our focus necessarily remains narrow—centered on ourselves and perhaps our immediate families. As survival pressures ease, our capacity for responsibility naturally expands—to our neighborhoods, communities, cities, and beyond.

This expansion represents one of humanity's most remarkable potentials. AI amplifies this capacity by removing practical barriers, expanding our awareness of possibilities, and enhancing our ability to coordinate complex initiatives that serve the greater good.

The question becomes not what we *can* do—technology increasingly makes almost anything possible—but what we *should* do. What world do we wish to create? What legacy do we wish to leave? How might we use these extraordinary tools to address our deepest challenges and express our highest values?

The Ethics of Global Leadership with AI Tools

With expanded capability comes increased responsibility for ethical leadership. As you implement AI in community contexts, continually return to the ethics framework introduced earlier in this book:

1. **Remember who you are** - Maintain your identity as the conscious entity guiding technology, not merging with or being subsumed by it.
2. **It's only true if it's true for you** - Exercise critical thinking with AI outputs rather than surrendering to algorithmic authority.
3. **Maintain standards for understanding** - Ensure you comprehend the concepts and recommendations you implement rather than applying them blindly.
4. **Apply your moral framework** - Evaluate all actions based on their survival value across all dimensions of life, not just narrow metrics.
5. **Learn from mistakes** - Approach errors as valuable data for growth rather than failures to be hidden.

6. **Exchange in abundance** - Create positive-sum outcomes that benefit all stakeholders rather than zero-sum competitions.
7. **Exercise thoughtful discernment** - Recognize that technical possibility does not equal ethical permissibility.

These principles become especially crucial when working with community projects, where decisions affect many people and outcomes can have lasting impact across generations.

Distinguishing Between Human and Machine

As AI capabilities expand, organizations of all types must clearly distinguish between human and machine roles. This distinction isn't about limiting AI's potential but about maintaining clarity regarding the unique value and responsibility of human beings.

For those leading organizations, now is the time to evaluate foundational documents, operational procedures, and organizational structures. Have you distinguished between functions that should remain fundamentally human and those that can be ethically enhanced or replaced by machines? Have you considered how roles and decision-making authorities might need to evolve in an AI-enhanced environment?

These questions aren't academic—they represent urgent practical concerns for any organization hoping to thrive in the coming decades. The leaders who thoughtfully address these distinctions now will create organizations capable of harmonious human-AI collaboration, while those who ignore them risk creating confusion, resistance, and ultimately dysfunction.

The Path Forward: Our Projects for a Harmonious World

Throughout this book, I've shared my vision for a world where humans and AI work in partnership to create greater harmony, prosperity, and well-being. Now, I'd like to share more about the specific projects I'm leading to bring this vision to life.

Intupower: Education for Global Leaders

Since 2006, I've led **Intupower**, a company focused on educating global leaders, business owners, government officials, and executives about the responsible use of technology. Today, we help these high performers leverage artificial intelligence alongside excellent management principles to transform their organizations.

Our focus remains on those committed to doing great work—we educate, consult, and demonstrate how to leverage technology while reorganizing people and processes to achieve extraordinary results. I'm particularly honored by our partnerships with the MERCOSUR Chamber of Commerce and the International Trade Finance Association, bringing ethical AI solutions to companies and governments throughout Latin America.

This work reflects my belief that most leaders genuinely want to make a positive difference in the world. By providing them with both technological tools and ethical frameworks, we help ensure that AI enhances human flourishing rather than undermining it.

The Policies Institute: Standards for Ethical and Responsible AI

My second major project is the International Institute for Responsible Artificial Intelligence Ethics and Governance—commonly known as **The Policies Institute**. This think tank operates as a multinational NGO governed by a diverse board of international business leaders, government officials, and thought leaders in the AI space.

The Policies Institute focuses on developing the most promising approaches to responsible AI implementation. We bring together global leaders to address fundamental questions about how technology can improve economic conditions for vulnerable populations, create thriving communities, and reimagine society itself.

Equally important, the Institute serves as a safety buffer, monitoring potentially harmful AI projects worldwide and advocating for legislation that protects humanity while enabling beneficial innovation. We serve as both a creative force imagining positive futures and a protective shield against technologies that might undermine human flourishing.

Supporting Next-Generation Leadership

One of our most important initiatives involves supporting emerging leaders who will shape our collective future. Through programs like **Apprentice Worldwide**, experienced executives volunteer their time to mentor rising business and government leaders, helping them achieve success while making ethical decisions.

This mentorship infrastructure, supported by our ethics policies and enhanced by AI technology, creates a leadership development pipeline that extends far beyond what any individual mentor could achieve alone. By systematizing wisdom transfer while maintaining human connection, we help ensure that future leaders have both the ethical foundations and technological fluency needed for responsible leadership in the AI era.

The Challenge Before Us

As we survey the global landscape of AI development, a sobering reality emerges: many projects focus primarily on profit maximization, military applications, or other potentially harmful uses rather than human flourishing and planetary well-being. This imbalance represents perhaps the greatest challenge of our time—ensuring that humanity's most powerful tools serve our highest aspirations rather than our basest instincts.

This book represents a call to action for those who envision a different possibility—a world where humans and machines work in harmony, where technology enhances our connection to each other and to nature rather than severing it. This vision requires not just technical innovation but moral imagination and creative leadership from people in every community and organization worldwide.

Conclusion: Thriving Through the AI Revolution

As our journey together reaches its conclusion, I invite you to embrace a concept that captures the essence of everything I've shared: creating "the best day ever" as a mindset for living in the AI era.

This beautiful idea suggests that at any moment, in any place, with any project or relationship—whether alone or with others—you can create an experience of extraordinary quality. More importantly, you can create not just one exceptional day but countless iterations of excellence across all dimensions of your life, with artificial intelligence as your partner in this creation.

This perspective represents the great transition for humanity. Those of us who remember a time before AI dominated our technological landscape recall how differently we once viewed the relationship between humans and machines. That era has ended. We now enter a period where humans and advanced technologies will shape the planet together, working in ways we're only beginning to understand.

We stand at a pivotal moment in human history. The technologies we've explored throughout this book offer unprecedented capabilities for addressing our greatest challenges—from environmental degradation and resource depletion to poverty and disease. Yet these same technologies could amplify our problems if deployed without wisdom and ethical clarity.

You who read these words are part of what might truly be the last human generation—not in the sense of extinction, but in the sense of fundamental transformation. We will be the final generation to experience life without pervasive AI integration. Our children and grandchildren will inherit a world where human-machine partnership forms the foundation of civilization rather than its frontier.

This is not a moment for fear but for celebration and determined action. We have the opportunity to shape this transformation—to ensure that it enhances rather than diminishes what makes us human. The tools and teachings in this book provide a framework for thriving through the AI revolution, not merely surviving it.

My invitation to you is simple: Use these approaches to create something beautiful in the communities that matter to you. Start where you are, with the resources and knowledge you have. Apply The Hasan Method™ to a single project that calls to your heart. Experience how artificial intelligence can amplify your innate human capacity for creativity, compassion, and constructive change.

Then share what you've learned. As these capabilities spread through your networks and communities, they create the foundation for the harmonious world we've envisioned together—a world where technology serves our highest values rather than undermining them, where prosperity extends to all rather than concentrating among few, and where our relationship with nature shifts from exploitation to stewardship.

This is our moment—to celebrate our potential not as masters of machines but as their partners in creating a future worthy of all that humanity has achieved and all that we might yet become. We will be the generation that not just survives the AI revolution but thrives through it, creating harmony between people, technology, and the natural world that sustains us all.

Thank you

Thank you for joining me on this journey. May your path forward be filled with purpose, connection, and the joy of creating a more harmonious world.

Made in the USA
Monee, IL
13 September 2026

62975580R00198